AI for Absolute Beginners

Understanding and Using ChatGPT

Maurice Kinsey

Maurice Kinsey

Contact@OrionNexus.io

https://OrionNexus.io

First Edition: August 2025

ISBN-13: 978-1-969209-00-0

Printed in the United States of America

To my father,

THE gold standard.

You taught me discipline and commitment to excellence through both words and actions. Your legacy lives on through everyone you touched.

And to my circle of life,

Who brings me love and peace.

You inspire me to live my purpose and turn dreams into goals.

Contents

Preface

C onsider this: a single, thoughtful conversation might fundamentally alter the trajectory of your business, transform your professional capabilities, or revolutionize how you approach everyday challenges.

Since the advent of computing, humanity has pursued tools to amplify our intellect and expand our potential. Today, that pursuit culminates in systems like ChatGPT – not merely programs, but partners capable of understanding and responding in natural language.

I am Maurice Kinsey, founder of OrionNexus.io and a perpetual student of technology's profound impact on human endeavor. In my work, I have witnessed firsthand how transformative AI tools enable businesses to achieve unprecedented efficiency, professionals unlock new creative possibilities, and curious individuals discover capabilities they never knew they possessed. Yet I have also observed the hesitation, the intimidation, and the unfortunate resignation that leads many to believe artificial intelligence is somehow beyond their reach.

This book exists to dismantle that barrier. Here, no coding expertise or technical degree is required. Your sole qualifications are curiosity and a willingness to engage. I believe that true mastery emerges through practice rather than theory. Thus, each chapter emphasizes practical application: crafting persuasive communications, streamlining routine processes, generating innovative solutions for persistent challenges.

To the everyday learner seeking to make sense of our AI-transformed world, to the professional looking to integrate these tools meaningfully into your work, to the small business owner wondering how to harness this potential – this book was written for you. My aim is not merely to teach you to use AI tools, but to help you develop them as strategic partners in your work and thinking.

Your first meaningful conversation with artificial intelligence awaits.

Maurice Kinsey

About the Author

Maurice Kinsey is a U.S. Army Veteran and Founder & CEO of OrionNexus.io. With over 20 years of experience in enterprise technology – including cybersecurity, emerging technologies, and next-generation systems – he helps businesses and professionals integrate AI tools safely and effectively.

Maurice's unique background combining military discipline, deep technology expertise, and business acumen gives him a distinctive perspective on AI adoption. He believes that the most powerful technologies should be accessible to everyone, not just technical experts. This philosophy drives his work at OrionNexus.io, where he helps small businesses and professionals navigate the AI landscape with confidence.

Maurice holds an MBA and a BS in IT Management, along with industry-leading certifications including CAITL (Certified Artificial Intelligence Transformation Leader), PMP (Project Management Professional), CISM (Certified Information Security Manager), and CISA (Certified Information Systems Auditor). He is known for his ability to translate complex technological concepts into clear, actionable guidance that real people can implement immediately.

CHAPTER 1

What is AI? (The Basics)

~

Opening Story: Sarah's Digital Awakening

Sarah Martinez had been running her family's bakery for fifteen years. At 52, she prided herself on doing things the traditional way – handwritten order forms, a paper calendar on the wall, and recipes passed down through generations. But on a particularly chaotic Tuesday morning, everything fell apart.

Three wedding cake orders got mixed up. The new employee misread her handwriting. A corporate catering request came in while she was elbow-deep in dough, and by the time she called back, they'd gone with a competitor. As Sarah sat in her flour-dusted office that night, surrounded by crumpled papers and feeling overwhelmed, her daughter Emma walked in.

"Mom, you know there are tools that could help with this, right?"

"I don't have time to learn complicated computer programs," Sarah sighed.

"What if I told you that you could just talk to it, like you're talking to me? And it would understand what you need?"

That was Sarah's first introduction to AI – not as some futuristic robot or complex programming, but as a helpful assistant that could understand her words and help organize her business. Within three months, Sarah was using AI to manage orders, create marketing content, and even develop new recipe variations. But we're getting ahead of ourselves. Let's start where Sarah did – with understanding what AI actually is.

Lesson 1.1: AI Explained Simply – Your Smart Helper

Imagine you're teaching a child to recognize dogs. You don't explain the biological classification system or the evolutionary history of Canis familiaris. Instead, you point to different dogs – big ones, small ones, fluffy ones, sleek ones – and say, "That's a dog." Eventually, the child learns to recognize dogs they've never seen before.

Artificial Intelligence works remarkably similarly. It's a computer system that has been "taught" through examples to recognize patterns and make decisions that normally require human intelligence. Just as that child learned to identify dogs, AI systems learn from thousands or millions of examples to understand language, recognize images, or predict what might happen next.

The Three Pillars of AI Understanding

1. Pattern Recognition At its core, AI excels at finding patterns in information. When you type a message on your phone and it suggests the next word, that's AI recognizing patterns in how people typically write sentences. It's not "thinking" about grammar rules; it's noticed that after "Thank you very," people often write "much."

I once worked with a small retail shop owner named Marcus who was skeptical about AI. To help him understand, I showed him his own sales records. "You already use patterns," I told him. "You know that umbrella sales go up when it's cloudy. You stock more ice cream in July. AI does the same thing, just with much more data and much faster."

2. Learning from Experience Unlike traditional computer programs that follow rigid rules, AI systems improve through experience. Think of it like a chef perfecting a recipe. Each time they make the dish, they adjust based on feedback – a little more salt here, cooking it two minutes less there. AI systems adjust their "recipes" for making decisions based on whether they got things right or wrong in the past.

3. Probability, Not Certainty Here's something crucial that many people misunderstand: AI doesn't deal in absolutes. It deals in probabilities. When an AI system identifies a photo of your cat, it's really saying, "Based on what I've learned, there's a 97% chance this is a cat." This is why AI sometimes makes mistakes that seem obvious to us – it's making educated guesses, not following foolproof rules.

Real-World Examples You're Already Using

Let's look at AI systems you probably interact with daily without realizing it:

Your Email's Spam Filter Remember when you used to get dozens of "You've won a million dollars!" emails? Your spam filter is an AI system that learned from millions of examples of spam emails. It notices patterns – certain words, sender addresses, formatting styles – and makes a probability-based decision: spam or not spam. The fascinating part? It keeps learning. When you mark an email as spam or move something out of the spam folder, you're teaching it to be more accurate for your specific needs.

Netflix's "Because You Watched" Recommendations Netflix doesn't have a team of people analyzing your viewing habits and personally selecting shows for you. Instead, its AI system recognizes patterns across millions of users. It notices that people who enjoyed "The Great British Baking Show" also tend to like "Queer Eye" and "Schitt's Creek." It's not understanding the content – it doesn't know these shows are heartwarming. It just recognizes the pattern in viewing behavior.

Your Phone's Face Unlock This is pattern recognition at its finest. Your phone's AI has learned the unique patterns of your face – the distance between your eyes, the shape of your nose, the contours of your cheeks. It can recognize you with glasses, without glasses, in different lighting, even with a different hairstyle. It's learned what variations of "you" still count as you.

The "Aha!" Moment: AI as Your Digital Apprentice

Here's the mental shift that helped Sarah from our opening story: Stop thinking of AI as a mysterious, all-knowing entity. Instead, think of it as a highly eager apprentice who:

- Learns by watching thousands of examples
- Gets better with practice and feedback
- Works incredibly fast but isn't infallible
- Excels at specific tasks but needs clear direction

When Sarah understood this, she stopped being intimidated. She realized she didn't need to understand the technical details any more than she needed to understand how her oven's thermostat worked. She just needed to know what it could do for her.

Common Misconceptions Debunked

"AI is going to become conscious and take over" Current AI is like a very sophisticated calculator. It processes information and recognizes patterns, but it has no consciousness, no desires, no agenda. It's a tool, like a hammer or a blender – just one that works with information instead of nails or fruit.

"AI understands everything like a human" When ChatGPT writes you a poem about your dog, it doesn't understand dogs, poetry, or emotions. It's recognized patterns in how humans write poems about pets and is creating something that fits that pattern. It's mimicry at an incredibly sophisticated level, but it's still mimicry.

"You need to be technical to use AI" Sarah runs a successful bakery using AI tools, and she still calls her daughter for help when the WiFi acts up. Using AI today is like using a smartphone – you don't need to understand the technology, just what buttons to push.

Your Turn: Discovering Your Daily AI

Take a moment to think about your day yesterday. Write down every digital tool or service you used. Now, let's identify the AI:

- Did your email filter any messages?
- Did your map app suggest the fastest route?
- Did any app send you a notification at just the right time?
- Did your phone's photos app group pictures by person or location?
- Did any streaming service recommend content?
- Did your credit card company flag any unusual activity?

Most people are surprised to discover they interact with AI dozens of times daily. You're not starting from zero – you're already an AI user. This book will help you become an AI power user.

The Foundation for Everything Else

Understanding that AI is pattern recognition, not magic, is crucial for everything we'll learn next. When we get to ChatGPT specifically, you'll see it's following the same principles – recognizing patterns in language to predict what words should come next. But unlike your spam filter that makes a simple yes/no decision, ChatGPT can generate entire essays, stories, or solutions by predicting one word at a time based on patterns it learned from reading billions of sentences.

As Sarah told me after three months of using AI in her bakery, "Once I understood it was just a really fast pattern-finder, not some sci-fi mystery, everything clicked. It's like having an assistant who's read every business book ever written and can help me apply those lessons to my specific situation."

Lesson 1.2: A Super Quick History of AI – From Dreams to Daily Reality

To understand where we are with AI, it helps to know how we got here. Don't worry – I promise this won't be a dry technical lecture. Think of it more as the story of humanity's quest to create a thinking assistant, full of brilliant insights, embarrassing failures, and unexpected breakthroughs.

The Dream Begins (1950s): Can Machines Think?

Picture a dreary afternoon in Manchester, England, 1950. Alan Turing, the brilliant mathematician who helped crack the Nazi Enigma code, is pondering a question that sounds like science fiction: "Can machines think?"

Rather than getting tangled in philosophy about consciousness and souls, Turing proposed something practical. He suggested that if a machine could have a conversation so convincingly human that you couldn't tell it apart from a real person, then for all practical purposes, it could be considered intelligent. This became known as the Turing Test, and it set the stage for everything that followed.

The 1950s were wildly optimistic about AI. Researchers genuinely believed they'd create human-level artificial intelligence within 20 years. At a groundbreaking conference at Dartmouth College in 1956, the term "artificial intelligence" was officially coined. The attendees were like the Avengers of early computing – John McCarthy, Marvin Minsky, Claude Shannon – names that would become legendary in computer science.

The First AI Winter (1970s): Reality Bites

Here's what those early pioneers didn't anticipate: making computers "think" was exponentially harder than they imagined. The computers of the 1970s had less processing power than your current smart doorbell. They could follow rules – if X, then Y – but they couldn't learn or adapt.

Imagine trying to teach someone to recognize cats by writing out rules:

- "A cat has four legs" (but so does a dog)

- "A cat has pointy ears" (but so does a fox)
- "A cat says meow" (but what about silent cats?)

You'd need millions of rules, and you'd still miss edge cases. Funding dried up. Researchers moved to other fields. AI entered what we now call its first "winter" – a period where progress stalled and enthusiasm waned.

The Learning Revolution (1980s-1990s): What If We Let Computers Learn?

The breakthrough came when researchers stopped trying to program explicit rules and started letting computers learn from examples. This shift to "machine learning" changed everything.

Dr. Geoffrey Hinton, often called the "Godfather of AI," was working on something called neural networks – computer systems loosely inspired by how brain neurons connect. His colleagues thought he was wasting his time. "Geoffrey, why are you still working on that?" they'd ask at conferences. Neural networks required massive amounts of data and computing power that simply didn't exist yet.

But Hinton and a small group of believers persisted. They were building the foundations for a future they could envision but not yet create.

The Perfect Storm (2010s): When Everything Clicked

Three things happened simultaneously that transformed AI from an academic curiosity into a world-changing technology:

1. Big Data Explosion Suddenly, we had data. Massive amounts of it. Every photo uploaded to Facebook, every tweet, every Amazon purchase, every Google search – it all became training data. Remember our analogy about teaching a child to recognize dogs? Now imagine showing them millions of different dogs every second. That's what became possible.

2. GPU Revolution Graphics cards, originally designed to render video game explosions, turned out to be perfect for AI calculations. It was like discovering your blender makes an excellent paint mixer – unexpected but transformative. What once took weeks to compute now took hours.

3. Algorithm Breakthroughs Remember Dr. Hinton? His neural networks finally had the data and power they needed. In 2012, his team entered an image recognition competition and absolutely destroyed the competition. It was like watching a Formula 1 car enter a go-kart race. The AI winter was officially over.

The ChatGPT Moment (2020s): AI for Everyone

Which brings us to November 30, 2022 – a date that might go down in history alongside the launch of the iPhone. OpenAI released ChatGPT to the public, and everything changed overnight.

Within five days, a million people had tried it. Within two months, 100 million. It was the fastest-growing application in human history. Why? Because for the first time, powerful AI felt accessible to everyone. You didn't need to code. You didn't need to understand algorithms. You just had to type.

I remember showing ChatGPT to my 75-year-old uncle, a retired plumber who proudly declared he was "allergic to computers."

Within minutes, he was asking it to write birthday poems for his grandkids and explain how modern heat pumps work. "It's like talking to the smartest person I've ever met," he said, "except they never get impatient with my questions."

Why This History Matters for You

Understanding this journey helps explain three crucial things:

1. We're Still Early If AI development were a baseball game, we're maybe in the third inning. The rapid progress feels overwhelming, but we're still figuring out what's possible. You're not late to the party – you're arriving just as things get interesting.

2. Progress Isn't Linear AI development has been a roller coaster of breakthroughs and setbacks. There will be more surprises, both positive and challenging. The key is staying adaptable and curious.

3. The Pattern Continues Each breakthrough came from making AI more accessible. From room-sized computers to smartphones, from complex programming to natural conversation – the trend is always toward easier, more intuitive use. ChatGPT is just the latest step in making AI a tool for everyone, not just techies.

Your Historical Moment

You're living through the most significant technological shift since the internet. But unlike the early internet, which required learning HTML and dealing with dial-up modems, today's AI meets you where you are. You can start using it immediately, in your own language, for your own needs.

Sarah from our bakery? She's part of this history now. Every small business owner who uses AI to compete with large corporations, every student who gets personalized tutoring, every creative who breaks through writer's block – they're all part of this unfolding story.

As we move forward in this book, remember: you're not just learning to use a tool. You're participating in a transformation that future generations will read about. And the best part? You don't need to understand the complex history or technical details. You just need to understand how to make it work for you.

Lesson 1.3: Types of AI – Understanding the AI Family Tree

Let me tell you about two conversations that perfectly illustrate the types of AI we're dealing with today.

The first happened last month. My friend Jennifer was showing off her new Tesla to her teenage daughter. "Watch this," she said, activating the self-parking feature. The car smoothly backed into a tight parallel parking spot while Jennifer sat with her hands off the wheel. Her daughter yawned. "Big deal. When will it drive me to school while I sleep in the back?"

The second conversation was with my colleague Robert, a financial advisor. He'd just spent an hour with ChatGPT, which had helped him write a complex retirement planning guide, analyze market trends, and even create a personalized investment quiz for his clients. "This is incredible," he said. "It's like having a Harvard MBA intern who never sleeps. But sometimes I wonder – does it actually understand finance, or is it just really good at sounding like it does?"

Both conversations touch on a fundamental distinction in AI that everyone needs to understand: the difference between Narrow AI and General AI, and why this distinction matters for how you use these tools.

Narrow AI (ANI): The Master of One

Think of Narrow AI like a world-class specialist. Just as a heart surgeon is incredibly skilled at cardiac procedures but wouldn't attempt brain surgery, Narrow AI excels at specific tasks but can't transfer that expertise to unrelated areas.

Every AI system you interact with today is Narrow AI. Let's explore what this means through concrete examples:

Your Phone's Voice Assistant When you ask Siri or Google Assistant about the weather, it seems magical. It understands your words, checks the forecast, and responds in natural language. But here's the thing – it's actually several narrow AI systems working together:

- One system converts your speech to text
- Another understands the intent behind your words
- A third retrieves weather data
- A fourth converts the response back to speech

Each component is brilliant at its specific job but useless outside it. The speech recognition system can't check whether any more than a translator can perform surgery.

The Chess Champion That Can't Play Checkers In 1997, IBM's Deep Blue defeated world chess champion Garry Kasparov. Headlines screamed about machines conquering human intelligence. But here's what they didn't mention: Deep Blue couldn't play checkers. It couldn't even play a simplified version of chess. It was engineered exclusively for standard chess, with all its specific rules and patterns.

I once explained this to a group of business owners using this analogy: "Imagine hiring the world's best sushi chef for your restaurant. They can create extraordinary sushi that wins awards and delights customers. But if you ask them to make pizza, they're starting from scratch. That's Narrow AI – exceptional depth, limited breadth."

ChatGPT: The Illusion of General Intelligence This is where it gets interesting. ChatGPT feels different from other narrow AI because its specialty – language – is so broad. It can write poetry, explain quantum physics, draft business plans, and tell jokes. It seems like it understands everything.

But here's the crucial insight: ChatGPT is still a narrow AI. Its specialty just happens to be predicting what words should come next based on patterns in text. It's like a musician who can play any song by ear but can't actually read music or understand music theory. The output sounds right, but the underlying process is pattern matching, not true understanding.

Real-World Example: The Medical AI Paradox

Dr. Sarah Chen, an oncologist I interviewed, shared a perfect example. Her hospital uses an AI system that can detect certain cancers in medical images with higher accuracy than most radiologists. "It saved lives," she told me. "It catches things human eyes miss."

But then she added something crucial: "Last week, a patient asked if the AI could look at her knee X-ray since she was already here. I had to explain that our cancer-detecting AI can't identify a broken bone. It would be like asking a wine expert to judge coffee – they're both beverages, but the expertise doesn't transfer."

This is the power and limitation of narrow AI. Within its domain, it can exceed human performance. Outside its domain, it's helpless.

General AI (AGI): The Science Fiction Dream

General AI is what most people picture when they think of AI – a system with human-like intelligence that can understand, learn, and apply knowledge across any domain. It's the AI from movies: Data from Star Trek, JARVIS from Iron Man, HAL from 2001: A Space Odyssey.

Here's the reality check: AGI doesn't exist yet, and experts disagree wildly on when (or if) it will arrive. Predictions range from 10 years to 100 years to never.

Why AGI Is So Hard Creating AGI isn't just about making narrow AI systems better. It's a fundamentally different challenge. Consider what human intelligence involves:

- Understanding context and nuance
- Applying knowledge from one area to solve problems in another
- Having genuine comprehension, not just pattern matching
- Possessing consciousness and self-awareness (maybe)
- Exhibiting creativity and original thought

I like to explain it this way: If narrow AI is like building really good tools – hammers, saws, drills – then AGI is like creating a master craftsperson who knows when and how to use each tool, can improvise new solutions, and understands why they're building what they're building.

Why This Distinction Matters for You

Understanding that all current AI is narrow AI helps you use it more effectively and avoid common pitfalls:

1. Set Appropriate Expectations When Jennifer's daughter expects their Tesla to become a fully autonomous chauffeur, she's expecting AGI behavior from a narrow AI system. The car's self-parking feature is remarkable, but it's solving a specific, well-defined problem in a controlled environment.

Similarly, when you use ChatGPT, remember it's a language pattern expert, not an all-knowing oracle. It can help you write and think through problems, but it's not actually understanding your business or your life.

2. Play to AI's Strengths Since narrow AI excels within its specialty, use it there. Don't ask your grammar-checking AI to do math. Don't expect your image recognition AI to write poetry. It's like hiring specialists – use them for what they're best at.

3. Maintain Human Oversight Because narrow AI lacks general understanding, it can make errors that seem bizarre to humans. A famous example: an image recognition AI classified a turtle as a rifle because of subtle pattern manipulations invisible to human eyes. The AI was following its training perfectly, but it lacked the general intelligence to think, "Turtles and rifles are fundamentally different categories of objects."

Practical Application: The Narrow AI Toolkit

Think of building your AI toolkit like assembling a team of specialists. Here's how different narrow AI tools might support someone like Robert, our financial advisor:

Writing and Communication: ChatGPT

- Client newsletters
- Email templates
- Educational content
- FAQ responses

Data Analysis: Specialized financial AI

- Market trend analysis
- Risk assessment
- Portfolio optimization
- Pattern recognition in financial data

Visual Content: Image generation AI

- Infographics
- Social media content
- Presentation visuals

Administrative: Scheduling AI

- Client appointment booking
- Calendar optimization
- Follow-up reminders

Each tool excels in its narrow domain. The key is knowing which tool to use when, and always applying human judgment to the output.

The Bottom Line: Embracing Narrow AI

Here's what I tell everyone who's starting their AI journey: Stop waiting for AGI. The narrow AI available today is already transformative if you use it correctly. It's like having access to a team of world-class specialists who work 24/7, never get tired, and constantly improve.

Sarah from our bakery doesn't need AGI. She needs narrow AI that can help with specific tasks: writing marketing copy, managing inventory, scheduling staff, and analyzing sales patterns. By understanding that each AI tool is a specialist, not a generalist, she can build a powerful support system for her business.

As we progress through this book, we'll focus on narrow AI – specifically ChatGPT and similar tools. You'll learn to leverage their strengths while understanding their limitations. Remember: you don't need artificial general intelligence to transform your work and life. You just need to get really good at using the remarkable narrow AI we have today.

The future might bring AGI that can do everything. But the present offers narrow AI that can help you do specific things better than ever before. Let's focus on mastering what's here now, not waiting for what might come later.

Lesson 1.4: Why AI is a Big Deal Now – The Perfect Storm

Three years ago, I was having coffee with Marcus, who runs a small marketing agency. He was frustrated, overwhelmed, and considering closing shop. "I can't compete anymore," he said. "The big agencies have teams of copywriters, data analysts, designers. I'm just one guy with two employees."

Fast forward to last month. Marcus's agency is thriving, he's hired three more people, and his client base has doubled. What changed? "AI leveled the playing field," he told me. "I now have access to capabilities that only enterprise companies could afford before. It's like going from fighting with a slingshot to having Iron Man's suit."

Marcus's transformation illustrates why AI has suddenly become the biggest technological shift of our time. It's not just one breakthrough – it's three massive changes happening simultaneously, creating what I call "The Perfect Storm" of AI advancement.

Factor 1: The Big Data Explosion – AI's Fuel

Remember our earlier analogy about AI learning like a child? Well, imagine trying to teach that child to recognize dogs by showing them only three pictures. They might learn that dogs are "brown and medium-sized," completely missing poodles, Great Danes, and chihuahuas. That was AI's challenge for decades – not enough examples to learn from.

Then everything changed. Let me paint you a picture of just how dramatic this data explosion has been:

The Numbers That Broke My Brain

- Every minute, people upload 500 hours of video to YouTube
- Every day, we send 347 billion emails
- Every second, Google processes 99,000 searches
- By 2025, we'll create 463 exabytes of data daily (that's 463 billion gigabytes)

But here's what makes this meaningful for AI: it's not just quantity, it's diversity. AI systems can now learn from:

- Billions of conversations in hundreds of languages
- Medical records from millions of patients
- Centuries of digitized books and articles
- Every possible way humans express ideas, solve problems, and create art

Real-World Impact: The Radiologist's New Partner Dr. Elena Vasquez, a radiologist in Miami, shared how data changed her field. "Our AI system has learned from 10 million mammograms. No human radiologist could see that many in a thousand lifetimes. It spots patterns we never knew existed."

But she added something crucial: "The AI doesn't replace me. It's like having a colleague who's seen every possible case. I make the final call, but now I make it with superhuman pattern recognition backing me up."

Factor 2: The GPU Revolution – AI's Engine

This part of the story involves an accident that changed everything. Graphics Processing Units (GPUs) were created for one purpose: making video games look amazing. They needed to calculate millions of pixels simultaneously to render those gorgeous gaming environments.

Then, around 2012, researchers had an "aha" moment. The math needed for graphics – parallel processing of massive calculations – was exactly what AI neural networks required. It was like discovering your gaming console was secretly a supercomputer.

The Speed Difference Is Mind-Blowing Let me put this in perspective. Training a modern language model on a traditional CPU (the regular computer processor) would take approximately 350 years. With GPUs, it takes a few months. It's the difference between walking to the moon and taking a rocket.

I witnessed this firsthand when visiting a tech startup in 2019. They showed me two computers trying to analyze customer sentiment from thousands of reviews:

- Traditional CPU: 6 hours and still running
- GPU-powered system: 3 minutes and done

The founder grinned: "Three years ago, this analysis would have cost us $50,000 in computing time. Now it costs $50."

The Democratization Effect Here's where it gets exciting for regular users. Cloud computing means you don't need to own these powerful GPUs. When you use ChatGPT, you're tapping into massive GPU clusters without buying any hardware. It's like having access to a Formula 1 race car through a ride-sharing app.

Marcus's small agency? He's using the same GPU-powered AI infrastructure as Fortune 500 companies. The playing field hasn't just leveled – it's been completely reimagined.

Factor 3: The Algorithm Breakthrough – AI's Brain

Imagine having all the ingredients for a gourmet meal and a perfect kitchen, but only knowing how to make toast. That was AI before the algorithm breakthroughs of the last decade. We had data, we had computing power, but we didn't know how to combine them effectively.

The game-changer came with something called the Transformer architecture in 2017. (Yes, like the robots, but much nerdier.) Without diving into technical details, here's what you need to know: Transformers allowed AI to understand context in a way it never could before.

The Context Revolution Before Transformers:

- AI read text word by word, like reading through a straw
- "Bank" could mean financial institution or river's edge, and AI struggled to tell which
- Connections between ideas got lost

After Transformers:

- AI can consider entire passages at once
- It understands that "I went to the bank to deposit money" is about finance
- It can maintain context across long conversations

Real-World Magic: The Lawyer's Story Patricia, a solo practice lawyer, shared her experience: "I was reviewing a 200-page contract. The AI not only found every mention of liability clauses but understood how they related to each other across different sections. It caught contradictions I missed after three readings. It's like having a photographic memory with legal training."

The Convergence: Why NOW Is Different

These three factors – data, computing power, and algorithms – converged around 2020, creating capabilities we couldn't imagine before. But the real catalyst was COVID-19.

The Pandemic Accelerator When the world went remote, digital transformation jumped from "nice to have" to "survive or die." Companies that planned five-year digital strategies implemented them in five weeks. Everyone from grandparents to grade-schoolers became comfortable with digital tools.

This created the perfect moment for AI to emerge from research labs into everyday life. People were ready for new tools, comfortable with digital interfaces, and desperately needed help managing increased workload and complexity.

The ChatGPT Moment When OpenAI released ChatGPT in November 2022, it wasn't just launching a product. It was lighting a match in a room full of gasoline. All the conditions were perfect:

- People were digitally literate from pandemic adaptations

- The technology had reached a usability tipping point

- The need for AI assistance was universal and urgent

What This Means for You: The Opportunity Window

Here's why understanding this perfect storm matters: We're in a unique historical moment. It's like being in California during the Gold Rush, except the gold regenerates every day and everyone can have a pickaxe.

The Early Adopter Advantage Studies show that businesses adopting AI now see average productivity gains of 40%. But here's the kicker – we're still in the early stages. Most of your competitors are still trying to understand what AI is, let alone using it effectively.

Remember Marcus? His success came not from being a tech genius but from being willing to learn and experiment while others hesitated. He started using AI for:

- First drafts of client proposals (saving 3 hours per proposal)
- Social media content creation (from 10 hours/week to 2)
- Data analysis that would have required hiring a specialist
- Customer service responses that maintain his agency's voice

The Compound Effect Each AI capability builds on others. When Marcus freed up time from writing first drafts, he could focus on strategy. Better strategy meant better results. Better results meant more clients. More clients meant more data to feed back into AI systems for even better insights.

This is happening across every industry. The question isn't whether AI will transform your field – it's whether you'll be driving that transformation or watching from the sidelines.

Your Action Plan: Riding the Wave

Understanding why AI is exploding now gives you the context to move forward confidently. You're not randomly jumping on a tech trend. You're positioning yourself at the convergence of three massive technological shifts that created this moment.

Here's how to think about it:

1. **Data is everywhere**: Your experiences, expertise, and unique perspective become valuable when combined with AI

2. **Computing is accessible**: You don't need technical skills or expensive hardware

3. **Algorithms are user-friendly**: Natural language interfaces mean you already have the skills you need

The perfect storm has created perfect conditions for anyone – regardless of technical background – to harness AI's power. The question isn't whether you should start using AI. The question is: What will you create with it?

As we dive into ChatGPT specifically in the next Chapter, remember: you're not just learning a tool. You're positioning yourself at the forefront of the biggest technological shift since the internet. And unlike the early internet, you don't need to learn to code. You just need to learn to talk to AI like the powerful assistant it can be.

Lesson 1.5: AI All Around You – The Invisible Revolution

Last Sunday, I watched my neighbor Tom prepare for a cross-country trip to visit his grandson. In the span of 30 minutes, he:

- Asked his phone for the weather forecast along his route
- Had his credit card app alert him to set a travel notice
- Let his car's navigation system find the fastest route avoiding construction
- Got a notification from his airline app about checking in
- Received personalized podcast recommendations for the drive
- Had his security camera system confirm it was in "away mode"

When I pointed out he'd just interacted with at least six different AI systems, Tom looked at me like I'd told him he'd been speaking

French all day without realizing it. "I thought AI was that ChatGPT thing everyone's talking about," he said.

Tom's experience perfectly illustrates a profound truth: AI hasn't just arrived – it's already woven into the fabric of our daily lives so seamlessly that we don't even notice it. It's like electricity – invisible, essential, and everywhere.

The AI You Don't See: Your Daily Digital Butler

Think of AI as an invisible butler who's been quietly making your life easier for years. This butler doesn't wear a tuxedo or speak with a British accent. Instead, it works behind the scenes, anticipating your needs and smoothing out friction you didn't even know existed.

Morning: The AI Wake-Up Committee

Let's follow a typical morning to see this invisible revolution in action:

6:30 AM - Your phone alarm goes off. But it's not just any alarm – if you use a smart alarm app, AI analyzes your sleep patterns through the night and woke you during a lighter sleep phase within your set window, making wake-up less jarring.

6:45 AM - You check your email. Behind the scenes, AI has already filtered out spam, categorized messages, and even suggested quick replies. That "Sounds good!" response suggestion? That's AI analyzing the context of the conversation and your typical response patterns.

7:00 AM - You ask your smart speaker for the weather while making coffee. Natural language processing AI converts your sleepy mumble into text, understands your intent, fetches the data, and converts the response back to speech – all in under two seconds.

7:30 AM - You scroll through your news feed. AI algorithms have curated stories based on your reading history, the time you typically spend on certain topics, and even how quickly you scroll past things you're not interested in. Every scroll teaches it more about your preferences.

The Photography Revolution in Your Pocket

Remember when taking a good photo required understanding aperture, shutter speed, and ISO? Now you point and shoot, and the results often rival professional photography. That's not better hardware – that's AI.

When you take a photo with your smartphone, here's what happens in milliseconds:

- AI detects faces and ensures they're in focus
- It recognizes the scene (sunset, food, portrait) and adjusts settings accordingly
- It might take multiple shots and blend them for optimal lighting
- Night mode uses AI to brighten dark shots without flash
- Portrait mode uses AI to blur backgrounds artistically

My photographer friend Lisa initially resented this. "It's cheating," she said. But now she sees it differently: "AI handles the technical stuff so I can focus on composition and creativity. It's like having an expert assistant who never misses focus."

The AI Safety Net: Protection You Never Notice

Some of the most important AI in your life is the kind you hope never to notice – the systems protecting you from fraud, scams, and security threats.

Your Financial Guardian Every time you swipe your credit card, AI systems spring into action:

- Analyzing if this purchase fits your normal patterns
- Checking if the location makes sense given your recent history
- Comparing against millions of fraud patterns
- Making a decision in under 200 milliseconds

I experienced this firsthand last year. Within minutes of someone attempting to use my card number in Romania (while I was sitting in Virginia), my bank called. "Our systems noticed unusual activity," they said. The AI had recognized that buying electronics in Bucharest didn't match my typical Saturday afternoon coffee shop purchase.

The Email Defender Your email's spam filter is like a bouncer at an exclusive club, but instead of checking IDs, it's analyzing:

- Sender reputation and authentication
- Language patterns typical of scams
- Suspicious attachments or links
- Unusual urgency or requests

The average office worker would receive 120+ spam emails daily without these filters. Instead, maybe one or two slip through. That's AI saving you hours every week.

The Personalization Engine: AI That Knows You

Whether we realize it or not, we're surrounded by AI systems trying to understand and anticipate our preferences. This can feel creepy or convenient, depending on your perspective, but it's undeniably pervasive.

The Netflix Mind Reader Netflix's recommendation system is a masterclass in AI personalization. It considers:

- What you watch (obviously)
- When you watch (late-night preferences differ from Sunday afternoon choices)
- How long do you watch (did you binge or abandon after 10 minutes?)
- What you search for but don't watch
- Even which thumbnails you hover over

My sister Jane discovered this during lockdown. "I realized Netflix was showing me and my husband different thumbnail images for the same show," she said. "His had action scenes, mine had character close-ups. It knew what would catch each of our attention."

The Shopping Assistant Amazon's "Customers also bought" isn't random. It's AI analyzing millions of purchasing patterns to predict what you might need next. It's remarkably accurate – sometimes disturbingly so.

A friend recently joked, "Amazon recommended a specific guitar tuner the day after I started learning guitar. I hadn't searched for anything music-related on Amazon. It just... knew." (Likely from his YouTube guitar lesson views, interconnected ad networks, and browsing patterns – AI connecting dots across platforms.)

The Hidden Workplace Revolution

Even if you think your job doesn't involve AI, it probably does. Modern workplaces are full of AI systems operating quietly in the background:

The Smart Office

- Calendar apps that find meeting times across schedules
- Expense systems that categorize receipts automatically
- Customer service platforms that route inquiries to the right department
- Project management tools that predict deadline risks
- HR systems that screen resumes for relevant experience

Real Impact: The Teacher's Tale Monica, a high school English teacher, thought she worked in a low-tech environment. Then she counted her AI interactions:

- Plagiarism detection software checking student essays
- Grading assistance tools suggesting rubric scores
- Reading level analyzers helping match students with appropriate texts
- Attendance systems using facial recognition for security

- Learning management systems recommending resources based on student performance

"I realized I was collaborating with AI all day," she said. "It handles the administrative burden so I can focus on actually teaching."

Navigation and Transportation: AI as Your Co-Pilot

Perhaps nowhere is AI more seamlessly integrated than in how we move through the world.

The Magic of Modern Maps When you ask for directions, AI doesn't just plot a route. It:

- Analyzes real-time traffic from millions of phones
- Predicts future congestion based on historical patterns
- Considers accidents, construction, and events
- Learns your preferences (highways vs. back roads)
- Even factors in where you're likely to find parking

Last month, my GPS suggested a bizarre detour through a residential neighborhood. I almost ignored it but followed the suggestion. Later, I learned there had been a major accident on my usual route. The AI had detected the slowdown from other drivers' phones before traffic reports caught up.

The Ride-Sharing Revolution Every Uber or Lyft ride involves multiple AI systems:

- Demand prediction positioning drivers before you even request
- Dynamic pricing based on supply and demand patterns

- Route optimization for drivers
- ETA calculations considering countless variables
- Matching algorithms pairing you with the ideal driver

The Streaming Symphony: AI as Your Personal DJ

Music and video streaming services have transformed from digital jukeboxes into AI-powered personal entertainment consultants.

Spotify's Secret Sauce Spotify's Discover Weekly playlist feels like magic – 30 songs every Monday that seem handpicked by someone who knows your soul. It's actually AI analyzing:

- Your listening history
- Songs you skip vs. complete
- Playlists you create
- Similar users' preferences
- Audio characteristics of songs you like (tempo, key, energy)

My nephew, a music snob who prided himself on finding obscure bands, admitted: "Spotify now finds bands I love before I do. It's like having a friend with perfect taste and infinite time to search for music."

Making Peace with Pervasive AI

Understanding that AI is already everywhere helps reframe our relationship with it. You're not starting from zero – you're already an experienced AI user. You've just been unconscious of it.

This matters for three reasons:

1. Confidence Building If you can successfully interact with the AI in your phone's photo app, smart speaker, and navigation system, you can definitely handle ChatGPT. The interfaces are getting more natural, not more complex.

2. Demystification AI isn't some far-future technology. It's here, it's helpful, and you're already proving daily that it's not too complicated for "non-tech people."

3. Intentional Use Once you recognize AI's presence, you can use it more deliberately. Instead of passively benefiting from AI curation, you can actively leverage these tools for your goals.

Your AI Inventory

Take a moment this week to notice your AI interactions. Keep a simple log:

- What AI did you interact with?
- What problem did it solve?
- What would that task have been like without AI?

Most people are shocked by how much AI they already use successfully. You're not learning to use AI from scratch – you're learning to use it intentionally.

Tom, my neighbor from the beginning of this story, had his revelation moment: "I've been complaining that AI is too complicated while using it successfully dozens of times a day. Maybe this ChatGPT thing isn't as scary as I thought."

Exactly, Tom. Exactly.

As we move forward in this journey, remember: AI isn't invading your life – it's already there, quietly helping. Now it's time to turn that unconscious collaboration into a conscious partnership. You've been dancing with AI all along. This book just teaches you to lead.

CHAPTER 2

Introducing ChatGPT

~

Opening Story: The Email That Changed Everything

David Chen stared at his laptop screen, the cursor blinking mockingly in the empty email window. It was 11 PM on a Thursday, and he needed to send a delicate message to his biggest client about a project delay. After 15 years running his construction company, David could build anything – except the right words for difficult conversations.

"Just be honest," his wife suggested from the doorway. "You always figure it out."

But this time felt different. The client was already frustrated. The delay wasn't entirely David's fault, but explaining the complexities of supply chain issues and subcontractor scheduling without sounding like he was making excuses seemed impossible.

David had heard about ChatGPT from his daughter over dinner last week. "Dad, it's like having a writing assistant who never judges you and always has ideas," she'd said. He'd brushed it off then – another tech fad the kids were into.

But now, at 11 PM, with his business relationship hanging on the perfect email, David decided to give it a try.

"Help me write a professional email explaining a project delay to an important client," he typed into ChatGPT, feeling slightly ridiculous talking to a computer. "I need to be honest about the challenges while maintaining their confidence in our ability to complete the project."

What happened next changed David's entire perspective on AI. ChatGPT didn't just spit out a generic template. It asked clarifying questions, suggested different tones, and helped him craft a message that was professional, empathetic, and solution-focused. Within 20 minutes, David had sent an email he felt proud of.

The client's response the next morning? "David, thank you for the transparency and clear plan forward. This is why we continue to work with you – your professionalism even when things get challenging."

That email marked the beginning of David's journey from AI skeptic to AI advocate. But to understand why ChatGPT could help David when he needed it most, we need to understand what ChatGPT actually is.

Lesson 2.1: What is ChatGPT? – Your AI Writing Partner

Imagine having a writing partner who has read essentially everything ever written – every book, article, website, and document available on the internet up to 2021. This partner never gets tired, never judges your ideas, and can adapt their communication style from casual to academic to creative in seconds. That's essentially what ChatGPT is, but let me explain what's really happening under the hood.

Breaking Down the Name

Let's decode "ChatGPT" like a secret message:

- **Chat**: It converses with you in natural language

- **GPT**: Generative Pre-trained Transformer

That last part sounds like sci-fi technobabble, so let's translate:

- **Generative**: It creates new text rather than just searching existing content
- **Pre-trained**: It learned from massive amounts of text before you ever met it
- **Transformer**: The type of AI architecture (remember from Chapter 1?)

Put simply: ChatGPT is an AI that learned to write by reading billions of pages of text and can now generate human-like responses to whatever you ask.

The Library of Babel Meets a Speed Writer

Jorge Luis Borges wrote about a library containing every possible book. ChatGPT is like having access to that infinite library, plus a speed writer who can instantly synthesize information from all those books into exactly what you need.

When David asked for help with his email, ChatGPT didn't copy-paste from some template database. It understood the pattern of professional emails, the tone needed for delicate situations, and the structure that maintains relationships while delivering difficult news. It generated something new, tailored to David's specific situation.

How ChatGPT Actually "Thinks"

Here's where people get confused. ChatGPT doesn't think like humans do. It's not pondering your question, considering options, or having insights. Instead, it's doing something both simpler and more complex: predicting what words should come next based on patterns it learned.

Think of it like this:

1. You type: "Help me write a professional email about..."

2. ChatGPT thinks: "Based on the trillions of sentences I've seen, when someone asks for help writing a professional email about something, the most likely helpful response starts with..."

3. It generates text one word at a time, each word influenced by all the words before it

It's like the world's most sophisticated autocomplete, but instead of finishing your sentence, it can continue for paragraphs, maintaining context and coherence throughout.

The Remarkable Scope of ChatGPT's Abilities

During my research for this book, I kept a log of how different people used ChatGPT in a single week:

Monday: Sarah (our baker) used it to write Instagram captions for her daily specials **Tuesday**: A lawyer friend drafted contract summaries for clients **Wednesday**: A teacher created personalized study guides for students **Thursday**: A novelist broke through writer's block on chapter 12 **Friday**: A job seeker polished their resume and cover letter **Saturday**: A parent got help planning a dinosaur-themed birthday party **Sunday**: A hobbyist learned to code their first Python program

Same tool, vastly different applications. That's the power of a general-purpose language AI.

What ChatGPT Is NOT

Understanding ChatGPT's limitations is just as important as knowing its capabilities. Let me clear up common misconceptions:

It's Not a Search Engine. When you ask ChatGPT about the weather or current events, it can't check the internet (in the basic version). It only knows what was in its training data. Think of it as a brilliant friend with amnesia about everything after its training cutoff.

It's Not a Database ChatGPT doesn't store information like a filing cabinet. It learned patterns from text, but it doesn't have perfect recall of specific facts. It's more like a student who studied hard and can give you very good answers based on what they learned, but might occasionally mix things up.

It's Not Conscious This is crucial: ChatGPT has no awareness, feelings, or understanding in the way humans do. When it says "I understand your frustration," it's following patterns of empathetic responses it learned, not actually feeling empathy.

Real-World Magic: The Stories That Show the Power

The Entrepreneur's Breakthrough Lisa had a brilliant idea for a sustainable fashion startup but froze when trying to write her business plan. "I knew what I wanted to say but couldn't organize it professionally," she told me. She used ChatGPT as a thought partner:

- First, she brain-dumped all her ideas
- ChatGPT helped organize them into sections
- Together, they refined each section
- She added her unique insights and data
- Result: A polished business plan that helped her secure funding

"It wasn't that ChatGPT wrote my business plan," Lisa clarified. "It helped me write the business plan that was already in my head."

The Student's Study Revolution Marcus, a medical student, was drowning in information. "I'd read about the circulatory system for hours and still feel confused," he said. He started using ChatGPT to:

- Explain complex concepts in simple terms
- Create memorable analogies
- Generate practice questions
- Clarify connections between topics

His grades improved, but more importantly: "I actually understand now, not just memorize. ChatGPT became my 24/7 tutor who never gets frustrated when I ask the same question five different ways."

The Interface: Simpler Than Texting

One of ChatGPT's greatest strengths is its simplicity. There's no manual to read, no complicated commands to memorize. You just:

1. Type what you want in plain language

2. Press enter

3. Read the response

4. Continue the conversation or start fresh

It's designed to feel like texting with a helpful friend. No technical knowledge required, no special syntax to learn. If you can send a text message, you can use ChatGPT.

The Evolution Happening Right Now

ChatGPT isn't static. It's evolving rapidly:

- **Newer versions** understand context better and make fewer mistakes

- **Specialized versions** can analyze images, create pictures, and browse the web

- **Custom versions** can be trained for specific industries or tasks

- **Integration capabilities** let it work within other tools you already use

When David first used ChatGPT for that crucial email, he accessed raw capability. Now, six months later, he has ChatGPT integrated into his email client, trained on his communication style, helping him maintain professional relationships with less stress.

Your ChatGPT Mindset

To use ChatGPT effectively, think of it as:

- **A writing partner**, not a replacement writer
- **A thought organizer** that helps structure your ideas
- **A creative catalyst** that offers perspectives you might not consider
- **A learning assistant** that explains things multiple ways
- **A productivity tool** that handles routine tasks so you can focus on what matters

But always remember:

- **You're the expert** in your life and work
- **You provide the judgment** about what's appropriate
- **You add the human touch** that makes communication meaningful
- **You verify important facts** before relying on them

Starting Your ChatGPT Journey

When David sent that successful email to his client, he didn't just solve a communication problem. He discovered a tool that would transform how he runs his business. Six months later, he uses ChatGPT for:

- Project proposals that win more bids
- Safety documentation that's clear and comprehensive
- Team communications that boost morale
- Customer service responses that build loyalty

- Marketing content that attracts ideal clients

"I'm still the same contractor," David says. "I just have a really smart assistant who helps me communicate as professionally as I build."

That's the essence of ChatGPT: It doesn't replace human intelligence and creativity. It amplifies them. It takes your expertise, your ideas, your unique perspective, and helps you express them more clearly, more professionally, more creatively than you might on your own.

As we dive deeper into how ChatGPT works and how to use it effectively, remember David's story. He didn't become a tech expert overnight. He just found a tool that helped him do what he already did well – serve his clients – even better.

Lesson 2.2: How ChatGPT "Thinks" – The Magic Behind the Curtain

My ten-year-old nephew Jake recently asked me, "Aunt Emma, is ChatGPT actually smart, or is it just pretending to be smart?"

I paused, realizing this was actually one of the most profound questions about AI I'd ever heard. The answer reveals something fundamental about ChatGPT that every user needs to understand: ChatGPT is simultaneously one of the most sophisticated technologies ever created and essentially a very elaborate pattern-matching system. It's not thinking – it's performing an incredibly convincing imitation of thinking.

Let me pull back the curtain and show you what's really happening when you chat with this AI.

The World's Most Advanced Autocomplete

Remember when you're typing a text message and your phone suggests the next word? "Thanks for..." might prompt suggestions like "your," "the," or "letting." That's autocomplete. Now imagine autocomplete on cosmic steroids:

- Instead of learning from your few thousand text messages, it learned from hundreds of billions of words

- Instead of suggesting one word, it can continue for thousands of words

- Instead of simple patterns, it learned incredibly complex relationships between ideas

That's essentially what ChatGPT is doing – predicting what word should come next, then the next, then the next, creating entire coherent responses one word at a time.

The Birthday Party Analogy

Here's how I explained it to Jake, and it's helped many adults understand too:

Imagine you're at hundreds of millions of birthday parties. You notice patterns:

- People usually sing before cake cutting

- The birthday person makes a wish before blowing out candles

- Guests say things like "Make a wish!" and "How old are you now?"

- Thank you speeches often include "I'm so grateful for..."

Now someone asks you to help plan a birthday party. You've never planned one yourself, but you've observed so many that you know exactly what typically happens and in what order. You can create a perfect birthday party plan by following the patterns you've observed.

ChatGPT has "attended" billions of conversations through its training. When you ask it something, it's drawing on all those patterns to construct a response that fits the patterns of helpful, accurate answers it has seen.

The Token Symphony

ChatGPT doesn't actually see words – it sees "tokens," which are chunks of text that might be whole words, parts of words, or punctuation. When you type "How do I make chocolate chip cookies?", ChatGPT sees something like:

- [How] [do] [I] [make] [chocolate] [chip] [cook] [ies] [?]

Then it starts its prediction symphony:

1. After seeing questions starting with "How do I make..." it knows responses often start with "To make..." or "Here's how to make..."

2. After "chocolate chip cookies," it predicts you'll want ingredients and steps

3. Each token it generates influences what comes next, maintaining coherence

The Context Window: ChatGPT's Working Memory

Imagine trying to have a conversation while only remembering the last few minutes. That's ChatGPT's context window – it can only "remember" a certain amount of the current conversation. For GPT-3.5, it's about 3,000 words. For GPT-4, it can be up to 25,000 words.

This explains some quirks:

- Why ChatGPT might forget something you mentioned earlier in a long conversation
- Why starting fresh sometimes gives better results
- Why being concise in your prompts leaves more room for detailed responses

Real-World Example: The Novelist's Dilemma Rebecca, a mystery writer, learned this the hard way. She was using ChatGPT to brainstorm plot ideas and kept adding details. After about 20 minutes of back-and-forth, ChatGPT started contradicting earlier plot points. "It forgot that the butler was already established as being in London during the murder," she laughed. Now she keeps a separate document with key plot points and starts fresh conversations for each chapter.

Probability, Not Certainty: The Dice Behind Every Word

Here's something that blows people's minds: ChatGPT isn't selecting the "correct" next word – it's choosing from probable next words. For any given point in text, there might be dozens of reasonable next words. ChatGPT calculates probabilities for each and makes a selection.

This is why:

- You can ask the same question twice and get different answers
- Sometimes it generates brilliant insights, sometimes nonsense
- The "temperature" setting in advanced uses controls how adventurous these choices are

Think of it like this: If ChatGPT is writing "The sky is...", it might calculate:

- "blue" - 40% probability
- "cloudy" - 20% probability
- "falling" - 0.01% probability
- "purple" - 2% probability

Usually, it picks high-probability options, creating sensible text. But that element of probability means it's always slightly unpredictable.

The Training Chronicles: How ChatGPT Learned to "Think"

Understanding ChatGPT's training helps explain both its capabilities and limitations. The process happened in stages:

Stage 1: The Great Reading ChatGPT read enormous amounts of text from the internet – websites, books, articles, forums. Imagine reading every book in every library, every article in every newspaper, every post on every forum. That's the scale we're talking about.

But here's the crucial part: It didn't memorize this text. It learned patterns. Like how after years of reading, you instinctively know that "once upon a time" is usually followed by a fairy tale, not a scientific paper.

Stage 2: Human Feedback Training This is where ChatGPT learned to be helpful rather than just coherent. Human trainers would:

- Ask questions and rate responses
- Provide examples of good answers
- Correct problematic outputs

It's like teaching a child manners – lots of "Yes, that's helpful!" and "No, we don't say that."

Stage 3: Reinforcement Learning ChatGPT learned to predict which responses humans would rate highly. This is why it tends to be helpful, harmless, and honest – it learned these get good ratings.

The Hallucination Problem: When Pattern Matching Goes Wrong

Because ChatGPT is pattern matching, not retrieving facts, it can confidently generate false information that sounds plausible. I call these "hallucinations," though really they're just patterns that don't match reality.

The Law Firm's Lesson A lawyer friend learned this dramatically. He asked ChatGPT for case law precedents and it generated three citations that sounded perfect – case names, years, jurisdictions. The only problem? None of them existed. ChatGPT had generated plausible-sounding legal citations based on patterns it had seen, not real cases.

"It was a wake-up call," he told me. "ChatGPT is brilliant for drafting and ideas, but I verify every factual claim."

Why Understanding This Matters

Knowing how ChatGPT "thinks" transforms how you use it:

1. Set Appropriate Expectations Don't expect perfect accuracy on facts – expect excellent pattern matching. Use it for:

- Creative writing and brainstorming
- Structuring and organizing ideas
- Explaining concepts in different ways
- Generating examples and analogies

2. Work With the Probability Nature If you don't like a response, try again. Ask differently. The probabilistic nature means variety is built in.

3. Provide Clear Context Since ChatGPT is pattern matching, give it clear patterns to match. "Write like a professional business consultant" gives it a specific pattern to follow.

4. Verify Important Information Always fact-check critical information. ChatGPT's confident tone doesn't indicate accuracy – it indicates pattern matching.

The Magic Is Real, Just Different

Jake, my nephew, eventually concluded: "So ChatGPT is like a really good mime? It acts like it's thinking but it's really just copying patterns?"

"Exactly," I told him. "But here's the thing – it's such a good mime that it can actually help you think better."

That's the paradox of ChatGPT. It doesn't truly understand anything, yet it can help you understand everything better. It doesn't have real intelligence, yet it can make you more intelligent in how you approach problems.

Practical Implications: Using ChatGPT's "Thinking" Effectively

Now that you understand the mechanism, here's how to work with it:

For Creative Tasks: Embrace the probabilistic nature. Generate multiple options. The pattern matching excels at creating variations.

For Analytical Tasks: Provide clear structure. "Let's think step by step" triggers patterns of logical reasoning.

For Learning: Ask for explanations in different styles. The pattern matching can present information in countless ways until one clicks.

For Writing: Use it as a sophisticated brainstorming partner. It's matching patterns of good writing, so let it suggest structures and approaches.

For Problem-Solving: Break complex problems into smaller patterns it can match. "First, let's identify the problem. Second, let's list possible causes."

The Bottom Line: A Thinking Tool, Not a Thinking Being

ChatGPT doesn't think – it performs an incredibly sophisticated imitation of thinking that's useful enough to transform how we work, learn, and create. It's like having access to a universal pattern library of human communication and being able to query it in natural language.

Understanding this doesn't diminish ChatGPT's usefulness – it enhances it. When you know you're working with the world's best pattern matcher, not an omniscient oracle, you can use it more effectively. You provide the genuine thinking, the context, the judgment. ChatGPT provides the patterns, the structure, the possibilities.

Together, you and this elaborate pattern-matching system can achieve things neither could alone. That's not artificial intelligence replacing human intelligence – that's artificial intelligence amplifying human intelligence.

And that's the real magic behind the curtain.

Lesson 2.3: Getting Started – Your First Conversation with ChatGPT

I still remember my friend Maria's first encounter with ChatGPT. She's a florist who prides herself on handwritten notes and personal touch. "I don't do computers," she'd always say. But when her daughter set her up with ChatGPT, Maria approached it like she was meeting a new employee.

"Hello," she typed carefully. "I need help writing a thank you note to a wedding client."

ChatGPT responded warmly, asked about the wedding's style, and helped her craft a note that made her tear up. "It sounds like me, but better," she said. "Like I had time to really think about what to say."

That's the beauty of getting started with ChatGPT – it's as simple as starting a conversation. But let me walk you through everything you need to know to make your first interaction as smooth as Maria's.

Step 1: Getting to ChatGPT

First, you need to get to ChatGPT. Here's the simplest path:

1. **Open your web browser** (Chrome, Safari, Firefox – any will work)

2. **Type**: chat.openai.com

3. **You'll see**: A clean, simple page with options to Log In or Sign Up

Think of this like walking into a helpful store. The door is always open, and there's no pressure to buy anything – there's even a free option.

Step 2: Creating Your Account

The sign-up process is refreshingly simple. OpenAI didn't clutter it with unnecessary steps:

1. **Click "Sign Up"**

2. **Enter your email** or use your Google/Microsoft account

3. **Create a password** (or skip this if using Google/Microsoft)

4. **Verify your email** (check your inbox for a confirmation link)

5. **Provide a name** (can be first name only)

6. **Verify you're human** (usually just clicking a box)

That's it. No lengthy forms, no credit card required for the free version.

Privacy Note: OpenAI can use free tier conversations to improve their models. If privacy is crucial, consider the paid version, which offers an opt-out option. But for learning and general use, the free tier is perfectly fine.

Step 3: Your First Screen

Once you're in, you'll see something beautifully simple:

- A text box at the bottom (where you type)
- Some example prompts in the middle (ignore these for now)
- A clean, uncluttered interface

It's designed to feel approachable, not intimidating. Like a blank page waiting for your thoughts.

Step 4: Your First Conversation

Here's where people often freeze. What do you say to an AI? Let me give you a secret: ChatGPT is incredibly forgiving. You can't break it, confuse it, or say the wrong thing. Start simple:

Great First Prompts:

- "Hello! Can you explain what you can help me with?"
- "I'm new to ChatGPT. What are you good at?"

- "Can you help me write a birthday message for my sister?"
- "Explain quantum physics like I'm five years old"
- "What's a good recipe for someone who can barely cook?"

The key is: just start. Type something, anything, and see what happens.

The Anatomy of a ChatGPT Conversation

Let me break down what happens when you interact:

1. **You type** your message in the box at the bottom
2. **You press Enter** (or click the send button)
3. **ChatGPT "thinks"** (you'll see dots animating)
4. **The response appears** progressively, like someone typing
5. **You can continue** the conversation or start fresh

Real Experience: Tom's First Chat Tom, a 67-year-old retiree, was nervous about trying ChatGPT. His first message: "I don't know what to ask."

ChatGPT responded: "That's perfectly okay! I'm here to help with all sorts of things. I could help you write something, answer questions, brainstorm ideas, or just have a friendly chat. What's on your mind today?"

Tom later told me, "It felt like talking to a patient librarian. No judgment, just helpfulness."

Essential Features to Know

The Regenerate Response Button Not happy with ChatGPT's answer? Click the regenerate button (usually circular arrows) and it'll try again with a different approach. Remember, ChatGPT is probabilistic – it can generate various responses to the same prompt.

Starting Fresh Sometimes you want a clean slate. Look for "New Chat" (usually in the sidebar or top). Each new chat starts with no memory of previous conversations.

Copying Responses Need to use ChatGPT's response elsewhere? Most interfaces have a copy button on each response. Click it to copy the entire text to your clipboard.

Editing Your Prompts Made a typo? Want to refine your question? You can often edit your previous messages and get new responses based on the correction.

Common First-Timer Mistakes (And How to Avoid Them)

Mistake 1: Being Too Formal "Dear ChatGPT, I hope this message finds you well. I am writing to inquire..." Better: "Hey, can you help me with..."

ChatGPT doesn't need formal greetings. Talk to it like a helpful colleague.

Mistake 2: Being Too Vague "Help me with work" Better: "I need to write an email declining a meeting invitation politely"

Specificity helps ChatGPT give you exactly what you need.

Mistake 3: Apologizing Constantly "Sorry to bother you, I know this is probably stupid, but..." Better: Just ask your question directly.

ChatGPT doesn't judge. It's here to help, period.

Mistake 4: Expecting Perfection Immediately ChatGPT might not nail it on the first try. That's normal. Refine your prompt, ask for adjustments, or try a different angle.

Privacy and Safety: What You Should Know

Before diving deep, understand these basics:

What ChatGPT Remembers

- Within a conversation: It remembers everything in that chat
- Between conversations: It remembers nothing
- About you personally: Only what you tell it in each conversation

Safe Sharing Guidelines

- Never share passwords or financial information
- Avoid sharing personally identifiable information unnecessarily
- Remember conversations may be reviewed for safety (in free tier)
- Treat it like a public conversation, not a private diary

Professional Use Considerations If using for work, check your company's AI policy. Some organizations have specific guidelines about what can be shared with AI tools.

Your First Week Game Plan

Here's a structured approach to getting comfortable:

Day 1: Basic introductions. Ask ChatGPT to explain something you're curious about.

Day 2: Try a practical task. Have it help write an email or text message.

Day 3: Explore creativity. Ask for a joke, short story, or poem.

Day 4: Test its teaching ability. Ask it to explain a concept you've always wondered about.

Day 5: Use it for planning. Ask for help organizing something – a trip, a meal plan, a workout routine.

Day 6: Dive into your interests. Discuss your hobby or profession.

Day 7: Reflection. Ask ChatGPT to help you summarize what you've learned about using it.

Real Success Stories: From First Chat to Daily Use

The Overwhelmed Parent Jessica, mom of three, first used ChatGPT at 2 AM when her baby wouldn't sleep. "I just typed 'my 6-month-old won't sleep help' in desperation." ChatGPT provided a calm, organized list of possibilities and gentle suggestions. "It was like having an experienced parent friend available 24/7."

The Small Business Owner Carlos runs a food truck. His first ChatGPT conversation: "I need to write a menu description for my new taco." Within minutes, he had five appetizing descriptions. "Now I use it for social media posts, event announcements, even recipe brainstorming."

The Nervous Student Amy, returning to college at 45, was intimidated by essay writing. Her first prompt: "I haven't written an essay in 20 years and I'm scared." ChatGPT walked her through essay structure, helped outline her thoughts, and built her confidence. "It never made me feel stupid for not knowing."

Troubleshooting Common Issues

"ChatGPT seems slow"

- Free tier can be slower during peak times
- Try refreshing the page
- Consider using it during off-peak hours

"I can't access it"

- Check your internet connection
- Try a different browser
- Clear your browser cache
- Make sure you're at chat.openai.com (not .com.ai or other variations)

"The responses seem generic"

- Add more detail to your prompts
- Ask for specific styles or approaches
- Use follow-up questions to refine

Your ChatGPT Starter Kit

Here are prompts to copy and try for different needs:

For Writing Help: "Help me write a [type of document] about [topic] for [audience]. The tone should be [formal/casual/friendly]."

For Learning: "Explain [concept] in simple terms. Give me an example to help me understand better."

For Problem-Solving: "I'm facing this challenge: [describe problem]. What are some potential solutions?"

For Creativity: "Give me 5 creative ideas for [specific need]."

For Daily Life: "Help me plan a [type of event/activity] for [number] people with [constraints/preferences]."

The Confidence Builder

Remember Maria, our florist from the beginning? A month after her first tentative "Hello," she was using ChatGPT to:

- Write product descriptions that increased sales
- Create care guides for different flowers
- Draft vendor emails in English (her second language)
- Brainstorm wedding flower combinations
- Plan seasonal promotions

"I thought AI was for tech people," she told me. "But it's just like training a really eager assistant. You just have to start talking to it."

Your Turn: Take the First Step

Right now, today, open ChatGPT and type something. Anything. Here are three prompts you can copy exactly:

1. "Hi! I'm brand new to ChatGPT. What are three interesting things you can help me with?"

2. "Can you help me write a friendly text message to reconnect with an old friend?"

3. "Explain why the sky is blue in a way that would make sense to a curious child."

The hardest part isn't learning to use ChatGPT – it's starting. Once you begin that first conversation, you'll discover what millions already know: it's not intimidating, it's empowering. It's not complicated, it's conversational. It's not replacing your intelligence, it's amplifying it.

Welcome to your journey with ChatGPT. The conversation starts now.

Lesson 2.4: Free vs. Paid Versions – Choosing Your ChatGPT Experience

When Rachel first discovered ChatGPT, she used the free version for three months and was thrilled. As a freelance graphic designer, it helped her write project proposals, brainstorm creative concepts, and communicate with clients more professionally. Then, during a critical client presentation prep at 9 AM on a Monday, she hit the dreaded message: "ChatGPT is at capacity right now."

"That was my wake-up call," Rachel told me. "I realized I was running a professional service on a free tool. It was like trying to run a design business with only the free version of design software – possible, but risky."

Rachel's story illustrates a crucial decision every ChatGPT user eventually faces: Is it time to upgrade? Let me break down everything you need to know about free versus paid ChatGPT, so you can make the right choice for your needs.

The Free Tier: More Powerful Than You Think

Let's start with good news: ChatGPT's free tier is genuinely useful, not a crippled trial version designed to force upgrades. OpenAI made a strategic decision to keep the free tier robust, and millions of users never need anything more.

What You Get for Free:

- Access to GPT-3.5 (highly capable, just not the absolute latest)
- Unlimited conversations (when available)
- Full conversational abilities
- No time limits on individual chats
- Access from any device
- Basic features like regenerating responses

Real User Perspective: The Teacher's Tale Mark, a high school history teacher, has used free ChatGPT for over a year. "I create lesson plans, generate discussion questions, and get help explaining complex historical events in different ways. I've never felt limited by the free version." His usage pattern – moderate, mostly during off-peak hours – makes free ChatGPT perfect for his needs.

Understanding the Limitations

However, free doesn't mean unlimited. Here are the constraints you'll encounter:

1. Availability Issues During peak times (typically weekday business hours in the US), free users might see "ChatGPT is at capacity" messages. It's like a popular restaurant – paying customers get reservations, walk-ins wait for available tables.

2. Older Model Version Free tier uses GPT-3.5, while paid users get GPT-4. The difference? Think of it as consulting a well-read graduate student (GPT-3.5) versus a seasoned professor (GPT-4). Both are helpful, but the professor has deeper insights and makes fewer mistakes.

3. No Advanced Features Free users miss out on:

- Web browsing (real-time information lookup)
- Advanced data analysis
- Image generation with DALL-E
- Custom GPTs (specialized versions)
- Plugin access

4. Response Speed During high-traffic periods, free tier responses can be notably slower. It's the difference between broadband and dial-up – both work, but one tests your patience.

ChatGPT Plus: The Premium Experience

At $20/month, ChatGPT Plus isn't just about removing limitations – it's about unlocking new capabilities. Here's what your subscription buys:

Always-On Availability Remember Rachel's presentation crisis? Plus users never see capacity messages. It's like having a 24/7 concierge versus hoping the front desk is staffed.

GPT-4 Access: The Intelligence Upgrade The difference between GPT-3.5 and GPT-4 is substantial:

- Better reasoning and logic
- More nuanced understanding
- Longer, more coherent responses
- Fewer factual errors
- Better at following complex instructions

Example: The Consultant's Comparison James, a business consultant, tested both versions with the same complex prompt about market analysis. GPT-3.5 gave a solid overview. GPT-4 provided a structured analysis with subsections, considered multiple perspectives, and even suggested potential blind spots in the analysis. "It was like the difference between a junior analyst and a senior partner," he said.

Cutting-Edge Features Plus subscribers get first access to new capabilities:

- **Web Browsing**: Ask about current events, research recent developments
- **Advanced Data Analysis**: Upload spreadsheets, create visualizations
- **DALL-E Integration**: Generate images from text descriptions
- **Custom GPTs**: Access specialized versions for specific tasks

The Power User's Secret: Custom GPTs

This deserves special attention. Plus users can access thousands of specialized GPTs created by the community, or build their own. Think of these as ChatGPT trained for specific jobs:

- **Legal Eagle GPT**: Specialized in legal document drafting
- **Code Mentor GPT**: Focused on programming help
- **Meal Planner GPT**: Creates shopping lists and recipes based on dietary needs
- **Academic Writer GPT**: Helps with research papers and citations

Success Story: The Real Estate Revolution Sandra, a real estate agent, uses a custom Real Estate GPT that knows property terminology, market trends, and listing best practices. "It's like having ChatGPT with a real estate license. My listing descriptions went from good to magazine-quality."

Making the Decision: A Practical Framework

Here's how to decide if Plus is worth it for you:

Upgrade to Plus If:

- You use ChatGPT for professional/business purposes
- You need reliable access during business hours
- You work with complex tasks requiring nuanced understanding
- Current events and real-time information matter
- You want to analyze data or generate images
- You're hitting free tier limits regularly
- $20/month is a reasonable business expense

Stay with Free If:

- You use ChatGPT casually or for learning
- You can work during off-peak hours
- Your tasks are straightforward
- You're just exploring AI capabilities
- Budget is a primary concern
- You rarely hit capacity limits

The ROI Calculator: Is Plus Worth It?

Let me share how different users calculate the value:

The Freelancer's Math "I charge $75/hour. If ChatGPT Plus saves me just 20 minutes per month, it pays for itself. It saves me hours." - Alex, freelance writer

The Small Business Perspective "One improved sales email that lands a client covers a year of ChatGPT Plus. It's the cheapest employee I've ever hired." - David, our contractor from earlier

The Student's Dilemma "$20/month is two hours at my campus job. But ChatGPT helps me understand concepts that would take much longer with just textbooks. It's better than a tutor at 1/10th the cost." - Emma, engineering student

Hidden Costs and Considerations

Before upgrading, consider these factors:

Subscription Fatigue Another $20/month subscription adds up. List your current subscriptions and honestly assess if you'll use ChatGPT enough to justify it.

Feature FOMO Don't upgrade just because features exist. Upgrade because you need those specific features. Web browsing sounds cool, but do you actually need real-time information for your use case?

The Learning Curve Advanced features require time to master. GPT-4 is more powerful but also more complex to prompt effectively. Budget time for learning.

Real-World Comparison: A Week in Both Worlds

I tracked my ChatGPT usage across both tiers for a week:

Monday (Free Tier)

- Morning: Smooth access, helped draft emails
- Afternoon: "At capacity" message, waited 15 minutes
- Evening: Slower responses but functional

Tuesday (Plus)

- Used GPT-4 for complex research summary
- Uploaded data spreadsheet for analysis
- Generated chart visualization
- No delays or interruptions

The difference wasn't just features – it was peace of mind and workflow consistency.

Tips for Each Tier
Maximizing Free ChatGPT:

1. Use during off-peak hours (evenings, weekends)

2. Save complex tasks for when you have time flexibility

3. Keep conversations focused to work within limits

4. Master prompt engineering to get better results from GPT-3.5

5. Have backup plans for capacity issues

Maximizing ChatGPT Plus:

1. Explore all features – you're paying for them

2. Try different Custom GPTs for specialized tasks

3. Use web browsing for current information

4. Leverage GPT-4 for complex reasoning tasks

5. Upload documents and data for analysis

6. Create your own Custom GPT for repeated tasks

The Hybrid Approach

Some users employ a clever strategy: they maintain both free and Plus accounts. They use free ChatGPT for routine tasks and switch to their Plus account for critical work. It's like having both a everyday car and a sports car – use the right tool for the journey.

Making Your Choice

Rachel, our graphic designer from the opening, has been a Plus subscriber for six months now. "The $20/month is invisible compared to the value. I land more clients, write better proposals, and never worry about availability during crucial moments."

But Mark, our history teacher, remains happily on the free tier. "My needs are met perfectly. Why pay for features I won't use?"

Both are right. The best choice depends on your specific situation, not what others do.

Your Next Step

Here's a practical way to decide:

1. **Use free ChatGPT exclusively for two weeks**
2. **Track every time you hit limitations**
3. **Note when advanced features would help**
4. **Calculate time lost to capacity issues**
5. **Estimate the value of that time**

If the value exceeds $20/month, upgrade. If not, enjoy the powerful free tool you have.

Remember: You can always upgrade later, and you can always downgrade. Start where you are, use what you have, and let your actual needs – not feature lists – guide your decision.

The beauty of ChatGPT isn't in having the most expensive version. It's in using whatever version you have to enhance your work, creativity, and life. Whether free or paid, you have access to transformative AI technology. The only wrong choice is not using it at all.

Lesson 2.5: Critical Limitations – What ChatGPT Can't (and Shouldn't) Do

Dr. Jennifer Huang was preparing for the most important presentation of her career – a funding pitch for her revolutionary cancer research. She'd been using ChatGPT to help structure her thoughts and polish her slides. The night before the presentation, she asked ChatGPT to fact-check a critical statistic about survival rates from a recent study.

ChatGPT responded confidently with specific numbers, even citing the journal and year. Jennifer almost included it in her presentation. But something felt off – the numbers seemed too perfect. She manually checked the source.

The journal was real. The year was correct. But the study didn't exist. ChatGPT had hallucinated the entire thing, complete with plausible-sounding statistics and a fictional research team.

"That was my wake-up call," Jennifer told me later. "ChatGPT is an incredible tool, but treating it as an infallible source could have destroyed my credibility. I learned to verify everything that matters."

Jennifer's near-miss illustrates why understanding ChatGPT's limitations isn't optional – it's essential for using it effectively and safely.

Limitation 1: The Hallucination Problem

ChatGPT's most dangerous limitation is its ability to generate false information that sounds completely plausible. It doesn't lie intentionally – it literally cannot distinguish between accurate and inaccurate information. It's pattern-matching, not fact-retrieving.

How Hallucinations Happen Remember, ChatGPT predicts what words should come next based on patterns. When asked about the capital of France, the pattern "capital of France is Paris" appears so frequently in its training that it reliably produces the correct answer. But for less common information, it might combine patterns in ways that create fiction.

Real-World Hallucination Examples:

- Inventing scientific studies with plausible titles and authors
- Creating historical events that never happened
- Generating fake quotes attributed to real people
- Making up features for software products
- Inventing legal precedents and case law

The Lawyer's Nightmare A now-infamous case involved lawyers submitting a brief with ChatGPT-generated citations. Every case cited was fictional. The lawyers faced sanctions, their client's case was damaged, and it became a cautionary tale across the legal profession. The fake cases had believable names, reasonable fact patterns, and proper citation format – but they didn't exist.

Protecting Yourself from Hallucinations:

1. **Verify critical facts** from authoritative sources

2. **Never use ChatGPT as your sole source** for important information

3. **Be especially cautious with**:

- Statistics and data

- Historical facts and dates
- Scientific findings
- Legal information
- Medical advice
- Recent events

4. Watch for hallucination red flags:

- Information that seems too convenient
- Perfect round numbers
- Extremely specific details about obscure topics
- Claims about recent events (beyond its training date)

Limitation 2: The Bias Inheritance

ChatGPT learned from the internet, which means it absorbed the internet's biases. Despite OpenAI's efforts to reduce bias, it still exists in subtle and not-so-subtle ways.

Types of Bias to Watch For:

- **Cultural bias**: Defaulting to Western/American perspectives
- **Language bias**: Better performance in English than other languages
- **Professional bias**: Assuming traditional career paths and roles
- **Historical bias**: Reflecting outdated societal norms from training data

The Entrepreneur's Discovery Amara, launching a business in Nigeria, noticed ChatGPT's business advice assumed American regulations, banking systems, and cultural norms. "It kept suggesting strategies that make no sense in Lagos," she said. "I learned to always specify my context and double-check cultural assumptions."

Working Around Bias:

- Be explicit about your context and location
- Challenge assumptions in ChatGPT's responses
- Ask for multiple perspectives on sensitive topics
- Use ChatGPT as a starting point, not final word
- Actively seek diverse sources to balance perspectives

Limitation 3: The Knowledge Cutoff

ChatGPT's training has a cutoff date. It doesn't know about events after that date and can't browse the internet (in the basic version). This creates a moving blind spot.

What This Means Practically:

- No awareness of recent news events
- Outdated information about rapidly changing fields
- No knowledge of new products, services, or companies
- Incorrect information about current political leaders
- Missing recent scientific discoveries or studies

The Investor's Mistake Robert asked ChatGPT about a hot tech startup to inform an investment decision. ChatGPT provided detailed analysis – based on the company's status two years ago. The startup had since pivoted entirely, changed leadership, and shifted markets. "I nearly made a decision based on ancient history," Robert reflected.

Handling the Knowledge Gap:

- Always verify current information elsewhere
- Include dates in your prompts when relevant
- For recent events, use ChatGPT Plus with web browsing
- Treat ChatGPT as a historical reference, not news source
- Cross-reference time-sensitive information

Limitation 4: No True Understanding

This is philosophical but practical: ChatGPT doesn't understand anything in the way humans do. It recognizes patterns and generates responses, but has no consciousness, emotions, or genuine comprehension.

Why This Matters:

- It can't truly empathize with your situation
- It doesn't understand consequences of its advice
- It can't judge ethical nuances in complex situations
- It has no personal experience to draw from
- It can't genuinely care about your wellbeing

The Therapist's Perspective Dr. Lisa Chen, a licensed therapist, uses ChatGPT for administrative tasks but warns against therapeutic use. "I've seen people pour their hearts out to ChatGPT. It responds with pattern-matched empathy, but there's no genuine understanding. It's like talking to a very sophisticated mirror – helpful for reflection, dangerous as a replacement for human connection."

Limitation 5: Context Window Constraints

ChatGPT can only "remember" a limited amount of conversation. Once you exceed this limit, it starts forgetting earlier parts of your chat.

Practical Impacts:

- Long documents get truncated
- Complex projects lose important details
- Earlier instructions get forgotten
- Contradictions appear in extended conversations
- Nuanced context gets lost

The Novelist's Workaround Marcus was writing a novel with ChatGPT's help. "By chapter 10, it forgot character details from chapter 1. I started keeping a separate 'story bible' document and beginning each session by reminding ChatGPT of key details."

Limitation 6: Inconsistent Reasoning

ChatGPT can be brilliantly logical one moment and absurdly illogical the next. Its reasoning abilities are pattern-based, not genuinely logical.

Examples of Reasoning Failures:

- Basic math errors (improved in GPT-4 but not eliminated)

- Logical contradictions within the same response
- Inability to truly count or track complex relationships
- Confusion with spatial reasoning
- Struggles with puzzles requiring genuine understanding

The Teacher's Test Professor Williams tested ChatGPT with logic puzzles from his philosophy class. "It solved some beautifully, then failed spectacularly on similar ones. It's like a student who memorized solutions without understanding principles."

Limitation 7: Security and Privacy Concerns

ChatGPT isn't a secure vault for sensitive information. Assume anything you type could potentially be seen by others.

Never Share:

- Passwords or login credentials
- Credit card or financial information
- Social Security numbers or government IDs
- Proprietary business information
- Confidential client data
- Private medical information
- Trade secrets or intellectual property

The Executive's Close Call A CEO almost pasted her company's entire strategic plan into ChatGPT for help with formatting. "I realized at the last second that I was about to share our most confidential information with an AI system. I created a version with fake company names and numbers instead."

Working Within the Limitations: Best Practices

Understanding limitations doesn't mean avoiding ChatGPT – it means using it wisely:

1. Trust but Verify

- Use ChatGPT for drafts, not final versions
- Fact-check anything important
- Cross-reference with authoritative sources

2. Layer Your Tools

- ChatGPT for ideation and structure
- Human expertise for accuracy and nuance
- Other tools for verification and current information

3. Set Appropriate Boundaries

- Professional assistance, not personal therapy
- Learning aid, not replacement teacher
- Writing partner, not sole author

4. Maintain Healthy Skepticism

- Question confident-sounding answers
- Ask for sources (then verify them)
- Get second opinions on critical decisions

5. Use Disclaimers When Sharing

- If sharing ChatGPT-assisted work, disclose it
- Never present ChatGPT's words as your expertise
- Be transparent about AI assistance

The Success Stories: Working With Limitations

The Smart Marketer Sarah runs marketing for a tech startup. "I use ChatGPT to brainstorm campaign ideas and write first drafts. But every fact gets verified, every claim gets sourced, and every piece gets human review. ChatGPT multiplies my productivity precisely because I know its limitations."

The Wise Student David uses ChatGPT throughout his engineering studies. "It's incredible for understanding concepts and working through problems. But I never trust its calculations without checking, and I always verify formulas. It's a study buddy, not an answer key."

The Careful Consultant Maya consults for Fortune 500 companies. "ChatGPT helps me structure presentations and explore angles I might miss. But client-specific information stays out, and every insight gets validated through traditional research. It's a thinking tool, not a thinking replacement."

Your Limitation Checklist

Before using ChatGPT for any important task, ask yourself:

- Does accuracy matter for this task?
- Am I sharing sensitive information?
- Do I need current information?

- Will I verify the output?

- Am I using it appropriately for the task?

- Do I have reasonable expectations?

The Empowering Truth

Here's the paradox: Understanding ChatGPT's limitations makes it more useful, not less. When you know exactly what it can and can't do, you can use it confidently within those boundaries.

Dr. Jennifer Huang, from our opening story, still uses ChatGPT daily. "That near-miss taught me to use ChatGPT as a brilliant assistant with specific blind spots. I wouldn't trust an intern with fact-checking critical data – why would I trust AI? But for brainstorming, writing, and thinking through problems? It's invaluable."

ChatGPT's limitations aren't bugs to be fixed – they're features to be understood. It's not a replacement for human intelligence, expertise, or judgment. It's a powerful tool that, used wisely, can amplify all three.

The key isn't avoiding the limitations – it's dancing with them. When you understand where ChatGPT fails, you can position it where it succeeds. And that's where the magic happens: not in pretending ChatGPT is perfect, but in perfectly understanding how to use this imperfect but powerful tool.

Remember: The most dangerous limitation isn't in ChatGPT – it's in treating it as limitless. Respect the boundaries, and within them, you'll find remarkable possibilities.

Lesson 2.6: Common Myths – Separating ChatGPT Fact from Fiction

Last month at a coffee shop, I overheard a fascinating conversation. Two business owners were debating whether to use ChatGPT:

"I heard it searches the entire internet in real-time," said one.

"That's nothing," replied the other. "My nephew says it's becoming self-aware. Like that movie with the robots."

A third person chimed in: "My professor says using it is cheating, period. No matter what."

I sat there thinking: three intelligent people, three complete myths about ChatGPT. These misconceptions aren't just wrong – they're preventing people from using an incredibly powerful tool effectively. Let's bust these myths once and for all.

Myth 1: "ChatGPT Searches the Internet in Real-Time"

This is probably the most common and problematic myth. People ask ChatGPT about yesterday's news, current stock prices, or recent sports scores, then get frustrated when it provides outdated or fictional information.

The Reality: Basic ChatGPT doesn't search anything. It's not connected to the internet at all. Instead, it generates responses based on patterns learned during training, which ended at a specific date. It's like asking a brilliant person who's been in a coma since 2021 about current events – they can only guess based on patterns they knew before.

Why This Myth Persists:

- ChatGPT responds confidently to current event questions
- It can make educated guesses that sometimes seem accurate
- People confuse it with search engines
- ChatGPT Plus does have web browsing, adding to confusion

Real Example: The Weather Fiasco Tom asked ChatGPT, "What's the weather in Seattle today?" ChatGPT responded with a plausible forecast: "Seattle is experiencing typical fall weather with temperatures in the mid-50s and a chance of rain."

Tom packed an umbrella. Seattle was having an unusual 78-degree sunny day.

"It wasn't lying," I explained to Tom later. "It was pattern-matching. Seattle plus fall usually equals rain. It gave you the statistical likelihood, not actual weather."

How to Work With This Reality:

- Never use basic ChatGPT for current information
- For recent events, use ChatGPT Plus with web browsing enabled
- Verify any time-sensitive information elsewhere
- Think of ChatGPT as a knowledgeable historian, not a news reporter

Myth 2: "ChatGPT Is Becoming Conscious/Self-Aware"

This myth ranges from exciting to terrifying, depending on who's telling it. Some people refuse to use ChatGPT because they think they're contributing to the robot uprising. Others expect it to have feelings and memory of past conversations.

The Reality: ChatGPT has exactly as much consciousness as your calculator – zero. It's a sophisticated pattern-matching system with no self-awareness, desires, goals, or feelings. When it says "I understand your frustration," it's following patterns of empathetic responses it learned, not actually understanding or feeling anything.

Why This Myth Persists:

- ChatGPT uses first-person language ("I think," "I understand")
- Its responses seem remarkably human-like
- Science fiction has primed us to expect AI consciousness
- The technology seems so advanced it must be aware

The Philosophy Professor's Test Dr. Sarah Mitchell teaches philosophy of mind. She regularly demonstrates ChatGPT's lack of consciousness to her students:

"I ask it about its subjective experiences – what it's like to be ChatGPT. It generates plausible-sounding responses about 'experiencing' text processing. Then I point out it's doing exactly what it does with everything else: pattern matching. It's seen thousands of descriptions of consciousness and is remixing them. There's no inner experience, just very convincing mimicry."

Why This Matters Practically:

- Don't feel guilty about "using" ChatGPT – it doesn't care
- Don't expect it to remember you between conversations

- Don't attribute malice to errors – there's no intent
- Don't worry about hurting its feelings
- Don't expect genuine understanding of your situation

Myth 3: "Using ChatGPT Is Always Cheating"

This myth is particularly harmful in educational and professional settings. Some people avoid a powerful tool entirely because they've heard it's inherently unethical to use.

The Reality: Using ChatGPT is like using any tool – it can be ethical or unethical depending on how you use it. Is using a calculator cheating at math? Is using spell-check cheating at writing? Context matters.

When It's Cheating:

- Submitting ChatGPT's work as entirely your own
- Using it when explicitly prohibited
- Bypassing learning objectives with shortcuts
- Misrepresenting your capabilities

When It's Not Cheating:

- Using it as a brainstorming partner
- Getting help with structure and organization
- Checking grammar and clarity
- Learning new concepts through dialogue
- Disclosed and appropriate use

The Teacher's Evolution Mrs. Rodriguez banned ChatGPT in her classroom initially. "I thought it would destroy student writing." Then she tried a different approach:

"I now teach students to use ChatGPT as a writing assistant, not a writer. They must submit their ChatGPT conversations along with their final essays. I grade their ability to prompt, evaluate, and improve upon AI suggestions. Writing skills evolved from just creating text to managing AI collaboration – a skill they'll need in any career."

Myth 4: "ChatGPT Knows Everything"

People often treat ChatGPT like an omniscient oracle, then feel betrayed when it makes obvious errors.

The Reality: ChatGPT knows patterns from its training data – nothing more, nothing less. It's remarkably broad but has significant gaps and inaccuracies. It's like a very well-read person who sometimes confuses facts or fills gaps with plausible-sounding fiction.

Knowledge Limitations Include:

- Information after its training cutoff
- Specialized or niche topics
- Personal or private information
- Real-time data
- Local knowledge
- Proprietary information

The Expert's Humbling Dr. James Park, a marine biologist, tested ChatGPT on his specialty – deep-sea volcanic vent ecosystems. "It got the basics right but invented species and mixed up research findings. It sounded authoritative while being subtly wrong. Anyone without expertise would be completely fooled."

Myth 5: "ChatGPT Will Replace Human Jobs Entirely"

This myth causes unnecessary panic and prevents people from learning valuable skills.

The Reality: ChatGPT is transforming jobs, not eliminating them. It's like how spreadsheets transformed accounting – accountants didn't disappear, they became more valuable by doing higher-level work.

Jobs Are Evolving:

- Writers now focus on strategy and editing AI-assisted drafts
- Programmers use AI to write boilerplate code faster
- Teachers use AI to personalize learning materials
- Marketers use AI for ideation and iterate on human creativity

The Copywriter's Transformation Lisa was terrified ChatGPT would end her career. "Then I realized it made me 5x more productive. I now handle strategy for five clients instead of just writing for one. My value shifted from word production to creative direction. I earn more and enjoy my work more."

Myth 6: "The Free Version Is Useless"

Some people believe you must pay for ChatGPT Plus to get any value.

The Reality: Free ChatGPT is incredibly powerful. Millions use nothing but the free version for:

- Daily writing tasks
- Learning and education
- Creative projects
- Problem-solving
- Basic coding help

The paid version adds capabilities, not core functionality.

The Bootstrapper's Success Marcus built his entire consulting business using free ChatGPT. "People said I needed Plus for 'serious' work. But free ChatGPT helped me write proposals, create frameworks, and develop training materials. I upgraded eventually, but only after free ChatGPT helped me afford it."

Myth 7: "ChatGPT Is Always Neutral and Unbiased"

People assume AI means objective, but that's dangerously wrong.

The Reality: ChatGPT inherited biases from its training data. While OpenAI works to reduce bias, it still exhibits:

- Cultural assumptions
- Language preferences
- Demographic biases
- Historical prejudices

- Regional perspectives

The Researcher's Discovery Dr. Aisha Patel studied ChatGPT's responses about different cultures. "Ask about 'professional attire' and it defaults to Western business wear. Ask about 'traditional medicine' and it shows skepticism rooted in Western scientific paradigms. It's not neutral – it's reflecting the biases in its training data."

How to Navigate Reality

Now that we've busted these myths, here's how to work with ChatGPT as it actually is:

1. Maintain Realistic Expectations

- It's a tool, not a magic solution
- It has significant capabilities and limitations
- It requires human judgment and verification

2. Use It Appropriately

- For appropriate tasks within its capabilities
- With proper disclosure when relevant
- As an assistant, not a replacement

3. Stay Informed

- Capabilities evolve rapidly
- New features appear regularly
- Best practices develop continuously

4. Think Critically

- Question its outputs
- Verify important information
- Understand its biases

The Liberation of Truth

Understanding what ChatGPT actually is – rather than what myths suggest – is liberating. You stop expecting magic and start leveraging reality. You stop fearing fiction and start using facts.

Rachel, a small business owner, put it perfectly: "Once I stopped believing ChatGPT was either useless or omnipotent, I found the sweet spot. It's like having a very smart intern who works 24/7, never gets tired, but sometimes confidently says ridiculous things. Perfect? No. Incredibly useful? Absolutely."

Your Myth-Free Action Plan

Starting today:

1. **List your assumptions** about ChatGPT
2. **Test them practically** with real tasks
3. **Note what works** and what doesn't
4. **Adjust your usage** based on reality, not myths
5. **Share accurate information** with others

The truth about ChatGPT is actually more interesting than the myths. It's not conscious, but it can help you think. It's not connected to the internet, but it contains patterns from vast human knowledge. It's not replacing humans, but it's augmenting human capabilities in unprecedented ways.

By seeing ChatGPT clearly – neither monster nor messiah, but a powerful tool with specific capabilities and limitations – you position yourself to use it effectively. The myths make ChatGPT either terrifying or disappointing. The reality makes it transformative.

Stop believing the coffee shop chatter. Start experiencing the actual capability. Your productivity, creativity, and success will thank you for it.

CHAPTER 3

The Art of Prompting

~~~

Opening Story: The $50,000 Prompt

Michael Torres had been using ChatGPT for months with mediocre results. As the owner of a struggling marketing agency, he'd tried using it for client proposals, but they always came out generic. "It's just not that useful," he'd complain to anyone who'd listen. "All the hype is overblown."

Then he attended a workshop where the instructor demonstrated two prompts:

Prompt 1 (What Michael had been using): "Write a marketing proposal"

Prompt 2: "Act as a senior marketing strategist with 15 years of experience in B2B SaaS companies. I need a proposal for a cloud storage startup targeting small law firms. Include their specific pain points around compliance and security. The proposal should be 2 pages, professional but conversational, and include 3 specific campaign ideas with metrics. Focus on trust-building and ROI."

The difference in outputs was staggering. The first produced generic marketing fluff. The second created a targeted, insightful proposal that addressed real client needs with specific, actionable strategies.

Michael rewrote all his prompts that week. Three months later, he landed his biggest client ever – a $50,000 contract. The CEO later told him, "Your proposal was the only one that truly understood our challenges."

"That workshop taught me something crucial," Michael reflects. "ChatGPT wasn't the problem. My prompts were. The difference between a lazy prompt and a crafted prompt isn't just better output – it's the difference between failure and success."

Lesson 3.1: What is a Prompt? – Your Instructions to the AI

Think of ChatGPT as the world's most capable assistant who also happens to be completely literal, has no ability to read your mind, and takes instructions exactly as given. The prompt is your instruction manual, your recipe, your blueprint for what you want. Master the prompt, master the output.

The Anatomy of Communication

At its simplest, a prompt is any text you type into ChatGPT. But that's like saying a painting is just paint on canvas. The art lies in how you structure that text to get what you want.

Let me show you the evolution of a prompt:

Level 1 - The Caveman Prompt: "sales email"

Level 2 - The Basic Prompt: "Write a sales email for my product"

Level 3 - The Detailed Prompt: "Write a sales email for my organic dog food product targeting health-conscious pet owners"

Level 4 - The Crafted Prompt: "Write a 150-word sales email for PurePaws, my organic dog food brand. Target health-conscious millennial pet owners who currently buy premium brands. Tone: friendly but professional. Include one customer testimonial, mention our grain-free formula, and end with a 20% first-order discount. Avoid pushy language."

Each level provides more context, more constraints, more direction. The result? Exponentially better outputs.

The Four Elements of Power Prompts

Through analyzing thousands of successful prompts, I've identified four elements that transform basic requests into power prompts:

1. Role Definition Tell ChatGPT who to be. This frames its entire response pattern.

- "Act as a experienced financial advisor..."
- "You are a creative writing teacher..."
- "Respond as a senior software engineer..."

2. Context Provision Give background information that shapes the response.

- "I'm a beginner learning Python..."
- "This is for a formal business presentation..."
- "My audience is 5th grade students..."

3. Specific Requirements State exactly what you want.

- "Include 3 examples"
- "Keep it under 200 words"
- "Use bullet points for key concepts"

4. Constraints and Style Define what you don't want and how it should sound.

- "Avoid technical jargon"
- "Use conversational tone"
- "No clichés or corporate speak"

The Psychology Behind Effective Prompts

Understanding why certain prompts work helps you craft better ones. ChatGPT responds to patterns, and certain patterns trigger more helpful responses:

Clarity Triggers Clarity When your prompt is crystal clear, ChatGPT doesn't have to guess. Ambiguity in equals ambiguity out.

Real Example: The Confusion Case Sarah asked: "Help with presentation" ChatGPT responded with generic presentation tips.

Sarah revised: "I'm presenting quarterly sales results to executives tomorrow. Help me structure 10 slides that emphasize our 25% growth while addressing the missed targets in the Northwest region. Suggest talking points for each slide." ChatGPT delivered a complete presentation outline with executive-appropriate messaging.

Specificity Unlocks Expertise The more specific your domain, the more ChatGPT can tap into relevant patterns.

Generic: "Marketing advice" Specific: "Instagram marketing strategies for a boutique yoga studio in Portland targeting women 25-40"

The specific version activates patterns related to Instagram, boutique fitness, Portland's culture, and that demographic's preferences.

The Prompt Spectrum: From Simple to Sophisticated

Not every task needs a complex prompt. Understanding when to use which level is key:

Simple Prompts for Simple Tasks:

- "Define photosynthesis"
- "Convert 100 euros to dollars"
- "List state capitals"

Medium Prompts for Structured Tasks:

- "Write a thank you email to a client after a successful project, mentioning future collaboration"
- "Explain blockchain to a small business owner considering accepting cryptocurrency"

Complex Prompts for Nuanced Tasks:

- "Analyze the pros and cons of remote work for a traditional manufacturing company. Consider employee morale, productivity metrics, cost implications, and company culture. Structure as a 1-page executive brief with recommendation."

The Art of Conversational Prompting

Here's a secret: you don't need to nail the perfect prompt immediately. ChatGPT excels at conversational refinement.

The Progressive Approach:

1. Start with basic prompt

2. Review output

3. Add clarifications

4. Request adjustments

5. Refine until satisfied

Real Conversation Example: User: "Help me write about productivity" ChatGPT: [Provides generic productivity tips] User: "Actually, I need a blog post about productivity specifically for night shift nurses" ChatGPT: [Adjusts to healthcare context] User: "Make it more personal, like advice from one nurse to another" ChatGPT: [Adopts peer-to-peer tone] User: "Perfect, but can you add a section about managing sleep schedules?" ChatGPT: [Adds requested section]

This iterative approach often yields better results than trying to craft the perfect prompt upfront.

Common Prompting Pitfalls

The Assumption Trap Assuming ChatGPT knows your context without providing it.

Bad: "Fix this" [pastes text] Good: "Proofread this email to my boss about requesting remote work privileges. Make it more persuasive while maintaining professionalism."

The Kitchen Sink Error Cramming too many requests into one prompt.

Bad: "Write a business plan and marketing strategy and financial projections and also make a logo design brief and suggest company names" Good: Break into separate prompts for each major task.

The Vague Direction Dilemma Using subjective terms without definition.

Bad: "Make it better" Good: "Improve clarity by simplifying complex sentences and adding transition phrases between paragraphs"

Prompt Templates That Work

Here are battle-tested templates you can adapt:

For Analysis: "Analyze [topic/document/situation] from the perspective of [role]. Focus on [specific aspects]. Provide [number] key insights and [number] actionable recommendations."

For Writing: "Write a [length] [type of content] about [topic] for [audience]. Tone should be [description]. Include [specific elements]. Avoid [things to exclude]."

For Learning: "Explain [concept] as if I'm [knowledge level]. Use [type of analogies/examples]. Break it down into [number] main points. Include a practical example of how this applies to [relevant situation]."

For Problem-Solving: "I'm facing [specific problem] in [context]. My constraints are [list constraints]. My goal is [desired outcome]. Suggest [number] potential solutions with pros and cons for each."

The Power of Priming

Advanced prompters use "priming" – setting context before the main request:

"I'm going to ask you to write marketing copy. Before we start, here's important context: My company sells eco-friendly water bottles to college students. Our brand voice is playful but informative. Our main differentiator is our lifetime warranty. Our customers care about sustainability but have limited budgets. Now, please write Instagram caption for our back-to-school campaign."

This context-setting dramatically improves output relevance.

Real-World Prompt Transformations

The Consultant's Evolution Before: "Business advice for my startup" After: "As a veteran business consultant specializing in B2C e-commerce, provide 5 specific strategies for reducing cart abandonment on my sustainable fashion website. Our current abandonment rate is 68%, mainly happening at shipping cost reveal. Budget is limited. Focus on quick wins implementable within 2 weeks."

Result: Actionable strategies that increased conversion by 15%.

The Teacher's Breakthrough Before: "Create a lesson plan" After: "Design a 45-minute lesson plan for teaching the American Revolution to 8th graders. Include a 10-minute engaging opener, 25-minute interactive main activity, and 10-minute assessment. Must accommodate visual and kinesthetic learners. Align with Common Core standards. No lecture-style teaching."

Result: An engaging lesson plan that became her most successful class of the year.

The Prompt Engineering Mindset

Effective prompting isn't about memorizing formulas – it's about developing a mindset:

1. **Think Like a Director**: You're directing a performance. Be specific about the role, scene, and desired outcome.

2. **Embrace Iteration**: Your first prompt is a rough draft. Refine based on results.

3. **Provide Context Generously**: When in doubt, give more background rather than less.

4. **Be Explicit About Format**: If you want bullets, say so. If you want paragraphs, specify.

5. **Define Success**: Make it clear what a good response looks like.

Your Prompting Action Plan

Starting today:

1. **Review your recent ChatGPT conversations.** Identify where vague prompts led to vague outputs.

2. **Rewrite three prompts** using the four elements (role, context, requirements, constraints).

3. **Create a prompt template** for your most common use case.

4. **Practice the conversational approach** – start simple and refine.

5. **Keep a "prompt journal"** of what works for your specific needs.

The Michael Torres Postscript

Remember Michael from our opening? He now teaches prompt engineering to his clients. "I realized that knowing how to prompt AI is like knowing how to Google effectively 20 years ago – it's becoming an essential skill. The difference between someone who can prompt and someone who can't isn't just efficiency. It's the ability to unlock possibilities."

Your prompts are your keys to ChatGPT's capabilities. Lazy prompts unlock little. Crafted prompts unlock everything. The beauty is, you don't need technical knowledge – just clarity about what you want and the willingness to communicate it effectively.

Master the prompt, and you master the machine. More importantly, you amplify your own capabilities in ways that would have seemed like magic just years ago. The prompt is your wand. It's time to learn the spells.

Lesson 3.2: The Golden Rules of Prompting – Getting Great Results

Emma Chen was brilliant at her job as a UX designer, but she had a problem. Every time she used ChatGPT to help with design documentation or user research synthesis, the outputs felt... off. Too generic. Too obvious. Missing the nuance that made her work special.

"I don't get it," she told me over coffee. "Everyone says ChatGPT is amazing, but it gives me the same bland stuff every time."

I asked to see her prompts. They were all variations of: "Write about user experience design" or "Create user personas."

"Let me show you something," I said, pulling up ChatGPT. "Let's apply the CLEAR framework."

Twenty minutes later, Emma was staring at outputs that captured design nuances she hadn't even thought to mention. "It's like it suddenly understands design thinking," she marveled.

"It always had that capability," I explained. "You just needed the right framework to unlock it."

Introducing the CLEAR Framework

After analyzing thousands of successful prompts and coaching hundreds of users, I've developed the CLEAR framework. It's not just another acronym – it's a systematic approach that transforms mediocre interactions into powerful collaborations.

C - Concise: Focused and efficient **L** - Logical: Structured thinking **E** - Explicit: Detailed specifications **A** - Adaptive: Iterative refinement **R** - Reflective: Learning from outputs

Let's dive deep into each element.

C - Concise: The Power of Focused Communication

Concise doesn't mean short – it means every word serves a purpose. It's the difference between a rambling request and a precision instrument.

The Problem with Word Soup Here's a real prompt I received from a client: "So I need help with this thing for my business where we're trying to figure out how to maybe get more customers but not just any customers but the right customers who will actually buy our premium services because we've been getting lots of leads but they're not converting and I think maybe our messaging is off or maybe it's the pricing or could be the website I'm not really sure what do you think we should do?"

ChatGPT's response was equally scattered – a little about messaging, some pricing advice, website tips. Everything and nothing.

The Concise Alternative "I run a B2B consulting firm. Problem: 100+ monthly leads but only 2% convert to our premium tier ($5K/month). Help me diagnose whether the issue is messaging, pricing, or qualification. Provide a systematic approach to identify the root cause."

This prompt is actually longer but far more concise – every detail serves the goal.

Conciseness Guidelines:

- One main objective per prompt
- Remove redundant information
- Include only context that affects the output
- Use specific numbers and details
- Eliminate hedge words ("maybe," "possibly," "kind of")

Real-World Conciseness Win Marcus, a financial advisor, transformed his prompts:

Before: "I need to write something about retirement planning for my clients who are getting older and worried about having enough money when they stop working, maybe include some stuff about social security and investments and healthcare costs"

After: "Write a 500-word retirement planning checklist for clients aged 55-65 with $500K-$1M in assets. Cover: optimal social security timing, healthcare cost planning, and tax-efficient withdrawal strategies. Tone: reassuring but action-oriented."

The focused prompt yielded a checklist that Marcus has used with dozens of clients.

L - Logical: Structured Thinking for Better Outputs

ChatGPT responds brilliantly to logical structure. When your prompts follow clear reasoning, the outputs mirror that clarity.

The Three-Part Logic Structure:

1. **Context** (Here's the situation)

2. **Challenge** (Here's the specific problem)

3. **Request** (Here's what I need)

Example of Logical Structure: "Context: I manage a 20-person remote software team across 3 time zones. Challenge: Our daily standups are inefficient – taking 45 minutes with low engagement. Request: Design a new standup format that takes 15 minutes maximum while improving team communication. Include specific agenda items and time allocations."

Logic Patterns That Work:

- **Sequential**: "First, analyze X. Then, based on that analysis, suggest Y."

- **Comparative**: "Compare approach A vs approach B for [specific criteria]."

- **Conditional**: "If [scenario 1], recommend X. If [scenario 2], recommend Y."

- **Hierarchical**: "Provide overview first, then break down into 3 main sections with 2-3 points each."

The Project Manager's Transformation Linda struggled with project summaries until she applied logical structure:

Illogical prompt: "Summarize project status"

Logical prompt: "Create executive project summary with this structure:

1. Current status (on track/delayed/at risk)

2. Key accomplishments this week (3 bullets)

3. Critical issues requiring decision (numbered list)

4. Next week's priorities (top 3) Keep total under 200 words."

Her CEO commented: "Your updates are the clearest on the team now."

E - Explicit: Detailed Specifications Eliminate Guesswork

Being explicit means spelling out exactly what you want. ChatGPT can't read your mind – but with explicit instructions, it doesn't need to.

The Explicit Checklist:

- **Format**: Paragraphs? Bullets? Table? Dialogue?
- **Length**: Word count? Number of points? Page length?
- **Tone**: Professional? Casual? Academic? Inspirational?
- **Audience**: Who will read/use this? What's their knowledge level?
- **Include**: What must be present?
- **Exclude**: What should be avoided?

Explicit vs Vague - Real Examples:

Vague: "Write about healthy eating"

Explicit: "Write a 300-word beginner's guide to Mediterranean diet. Audience: busy professionals aged 30-45. Include: 5 key foods to add, 3 foods to reduce, sample one-day meal plan. Exclude: complex recipes, expensive ingredients. Tone: encouraging and practical."

The Explicitness Multiplier Dr. Sarah Kim, a therapist, discovered that explicitness transformed her patient handouts:

"Instead of asking for 'stress management tips,' I now request: 'Create a one-page stress management handout for anxiety patients. Include: 3 breathing exercises with step-by-step instructions, 2 grounding techniques for panic moments, 1 daily prevention practice. Use simple language (8th-grade level), calming tone, and add brief scientific explanation for why each technique works. Format with clear headings and plenty of white space.'"

The result? Handouts that patients actually use and share.

A - Adaptive: Iterative Refinement

The magic of ChatGPT isn't in perfect first responses – it's in the ability to refine and adapt. Adaptive prompting means treating interactions as conversations, not one-shot commands.

The Adaptive Process:

1. **Initial prompt** (80% of what you want)

2. **Review output** (What's missing? What's off?)

3. **Refine request** ("Good start, but please adjust X")

4. **Iterate until optimal** (Usually 2-3 rounds)

Adaptive Phrases That Work:

- "That's helpful. Now make it more [specific quality]"
- "Good framework. Apply it to [specific situation]"
- "Keep the structure but adjust the tone to [description]"
- "Expand on point 3 and condense point 1"
- "Rewrite with [specific audience] in mind"

Real Adaptive Session: Watch how Jake, a startup founder, refined his investor pitch:

Prompt 1: "Write an investor pitch for my edtech startup" Output: Generic pitch

Prompt 2: "Make it specific to K-12 adaptive learning software for math" Output: Better but still generic

Prompt 3: "Focus on our unique AI algorithm that identifies learning gaps 3x faster than competitors. Add specific metrics: 10,000 users, 87% improvement rate" Output: Much stronger

Prompt 4: "Perfect structure. Now make it more story-driven – start with a struggling student named Maria" Output: Compelling pitch that helped secure funding

R - Reflective: Learning from Every Interaction

Reflective prompting means analyzing what worked, what didn't, and why. It's about building your personal prompt playbook.

The Reflection Questions:

- What specific phrases generated the best outputs?
- Which role definitions work for your needs?
- What level of detail gives optimal results?
- Which formats suit your use cases?

Building Your Prompt Library Susan, a marketing director, keeps a spreadsheet of successful prompts:

- Email templates prompt that captures her brand voice
- Blog post structure that ranks well
- Social media series framework
- Customer persona generator

"I don't reinvent the wheel anymore," she says. "I've got proven prompts for 80% of my tasks."

The CLEAR Framework in Action

Let me show you the complete framework applied to a real scenario:

Scenario: Restaurant owner needs a response to a negative review

Without CLEAR: "Help me respond to bad review"

With CLEAR: **C**oncise: "Draft response to negative Yelp review about slow service last Saturday" **L**ogical: "Context: Popular family restaurant, usually prompt service. Challenge: Unexpected large party caused delays. Request: Professional response that acknowledges issue and shows improvement" **E**xplicit: "100 words, apologetic but not groveling, mention our new reservation system, invite them back with personal touch" **Adaptive:** [After first draft] "Good, but make it warmer and add specific detail about our usual 15-minute serve time" **Reflective:** [Saves template for future review responses]

Advanced CLEAR Techniques

The Pre-Flight Check Before hitting enter, run through:

- C: Is every word necessary?
- L: Does the logic flow clearly?
- E: Have I specified format, length, tone, audience?
- A: Am I prepared to refine?
- R: Will I learn from this interaction?

The Power Combo Combine elements for exponential results: "[LOGICAL STRUCTURE] Create a three-part analysis of [CONCISE OBJECTIVE]. [EXPLICIT SPECS] Format as executive brief, 500 words max, professional tone, include 2 data points per section. [ADAPTIVE READY] I'll provide feedback for refinement."

Common CLEAR Violations and Fixes

Violation: "Make it better" **CLEAR Fix:** "Improve clarity by adding topic sentences to each paragraph and transitions between sections"

Violation: "Write something creative" **CLEAR Fix:** "Write a 200-word product description for eco-friendly notebooks using storytelling approach, highlighting sustainability journey from forest to desk"

Violation: Information dump without structure **CLEAR Fix**: Organize into logical sections with clear hierarchy

Your CLEAR Implementation Plan

Week 1: Focus on Concise

- Review old prompts and remove unnecessary words
- Practice one-objective prompts

Week 2: Add Logical

- Structure every prompt with context-challenge-request
- Experiment with sequential and comparative patterns

Week 3: Emphasize Explicit

- Create specification checklists for common tasks
- Practice being ultra-specific about format and tone

Week 4: Develop Adaptive skills

- Plan for 2-3 refinement rounds
- Build phrase library for adjustments

Ongoing: Reflective practice

- Keep prompt journal
- Build personal template library
- Share successes with others

The Emma Chen Epilogue

Remember Emma, our UX designer? Six months later, she's become the go-to person in her company for AI-assisted design documentation. "CLEAR changed everything," she told me. "But the real shift was realizing that prompting is a skill like any other. The more intentionally I practiced, the better I got."

Her latest project? Teaching CLEAR to her entire design team. "It's not about the framework itself," she explains. "It's about thinking clearly about what you want before asking for it. That's a skill that helps with or without AI."

Your CLEAR Advantage

The difference between frustration and flow with ChatGPT isn't the technology – it's the prompting. CLEAR gives you a systematic way to unlock ChatGPT's full potential.

But here's the secret: CLEAR isn't just about better AI outputs. It's about clearer thinking. When you learn to be Concise, Logical, Explicit, Adaptive, and Reflective in your prompts, you're developing skills that enhance all communication.

Master CLEAR, and you don't just get better at talking to AI. You get better at clarifying your own thoughts, structuring your ideas, and achieving your goals. That's the real golden rule of prompting: The clearer you are about what you want, the more likely you are to get it – from AI and from life.

Lesson 3.3: Hands-On Prompt Exercises – Practice Makes Progress

"I understand the theory," said David, a financial analyst attending my workshop. "But when I sit down at ChatGPT, my mind goes blank. It's like knowing how to swim in theory but freezing at the pool's edge."

I smiled. "Then let's get you in the water. No more theory – just practice."

What followed was two hours of hands-on exercises that transformed David from a hesitant prompter to someone crafting sophisticated queries. By the end, he was teaching other participants tricks he'd discovered.

"It's like learning to cook," David reflected later. "Recipes help, but you learn by doing, tasting, adjusting."

This lesson is your hands-on kitchen. We'll work through real exercises, analyze what makes prompts succeed or fail, and build your prompting muscles through deliberate practice.

Exercise Set 1: The Transformation Challenge

Let's start by taking weak prompts and making them powerful. For each exercise, try writing your improved version before reading mine.

Exercise 1.1: The Vague Request Weak prompt: "Help with presentation"

Try improving this yourself first...

Strong prompt: "I'm presenting quarterly sales results to C-suite executives in 3 days. Create an outline for a 15-minute presentation that celebrates 20% growth while addressing concerns about increased customer churn. Include talking points for 5 slides and potential tough questions they might ask."

What changed:

- Added specific context (quarterly sales, C-suite)
- Defined constraints (15 minutes, 3 days)
- Included both positives and challenges
- Requested specific deliverables (outline, talking points, questions)

Exercise 1.2: The Kitchen Sink Weak prompt: "I need marketing help and also sales strategies and maybe some customer service tips and brand ideas for my new business"

Your turn to improve...

Strong prompt: "I'm launching an online fitness coaching business next month. Priority 1: Create a 30-day pre-launch marketing plan focused on building an email list of 500 potential clients. Include daily actions, content themes, and 3 lead magnet ideas."

What changed:

- Focused on one priority instead of everything
- Added specific timeline and metrics
- Requested actionable output (daily actions)
- Provided clear context (online fitness coaching)

Exercise 1.3: The Missing Context Weak prompt: "Write an email about the meeting"

Improve this prompt...

Strong prompt: "Write a follow-up email to 5 team members after our product roadmap meeting. Summarize the 3 features we prioritized for Q2, assign action items discussed (with owners and deadlines), and set expectations for next Friday's check-in. Tone: professional but encouraging, around 200 words."

What changed:

- Specified email type (follow-up)
- Identified audience (5 team members)
- Added crucial details (product roadmap, Q2 features)
- Included structure and tone requirements

Exercise Set 2: Role-Playing for Better Results

Different roles unlock different capabilities in ChatGPT. Practice these role-based prompts:

Exercise 2.1: The Expert Consultant Task: Get advice on improving customer retention

Basic: "How do I keep customers?"

Role-enhanced: "Act as a customer retention expert who's worked with 100+ SaaS companies. My B2B software has 15% monthly churn (industry average is 10%). Diagnose likely causes and provide 5 specific retention strategies I can implement within our $10K budget and 2-person team."

Practice writing your own role-enhanced version for a different scenario...

Exercise 2.2: The Creative Professional Task: Name a new product

Basic: "Suggest names for my app"

Role-enhanced: "You're a brand naming specialist who's created names for successful tech startups. Generate 10 names for my productivity app that helps remote workers manage deep focus time. Names should be: memorable, suggest focus/flow, available as .com domains, and work globally. Avoid overused prefixes like 'Pro' or 'Smart'."

Exercise 2.3: The Teacher Task: Learn a complex concept

Basic: "Explain machine learning"

Role-enhanced: "You're a patient computer science professor known for brilliant analogies. Explain machine learning to me like I'm a small business owner with no technical background. Use a relatable analogy from running a restaurant, break it into 3 main concepts, and give me one practical example of how I could use it in my business."

Exercise Set 3: The Constraint Challenge

Constraints paradoxically increase creativity. Practice adding helpful limitations:

Exercise 3.1: Format Constraints Task: Summarize a business idea

Unconstrained: "Summarize my business idea"

Constrained: "Summarize my eco-friendly packaging business in exactly 3 formats:

1. One-sentence elevator pitch (max 20 words)

2. Three-bullet value proposition

3. 100-word paragraph for investors Focus on sustainability impact and cost savings for e-commerce brands."

Exercise 3.2: Style Constraints Task: Explain a policy change

Unconstrained: "Announce new remote work policy"

Constrained: "Write 3 versions of an email announcing our new hybrid work policy (3 days office, 2 days remote):

1. For executives: Focus on productivity and culture benefits (formal, 100 words)

2. For employees: Emphasize flexibility and work-life balance (friendly, 150 words)

3. For clients: Assure service continuity (professional, 75 words)"

Exercise Set 4: The Iteration Workout

Practice the art of progressive refinement:

Exercise 4.1: The Three-Round Refinement

Round 1: "Write a bio for my LinkedIn" Output: [Generic professional bio]

Round 2: "Make it specific to my role as a sustainability consultant in the fashion industry. Highlight my 10 years experience and focus on helping brands reduce environmental impact" Output: [More targeted but still formal]

Round 3: "Perfect background. Now make it more personable – add why I'm passionate about sustainable fashion and end with an invitation to connect. Keep professional but warm" Output: [Engaging bio that balances expertise with personality]

Try your own three-round refinement on a different topic...

Exercise Set 5: Chain-of-Thought Prompting

Advanced technique: Guide ChatGPT through logical steps:

Exercise 5.1: Problem-Solving Chain Instead of: "How do I increase website traffic?"

Try: "Let's think through increasing website traffic step by step:

1. First, identify my current traffic sources and volumes

2. Then, analyze which sources have highest conversion

3. Next, list 5 ways to amplify high-converting sources

4. Finally, create 30-day action plan

My site: B2B software blog, currently 10K monthly visitors, 60% from search, 30% social, 10% direct."

Exercise 5.2: Decision-Making Chain "Help me decide whether to hire a full-time developer or continue with freelancers. Let's analyze:

1. Calculate current freelance costs vs full-time salary

2. List pros/cons for my specific situation (early-stage startup, $2M funding, technical product)

3. Consider impact on company culture and product development

4. Provide recommendation with reasoning"

Real-World Practice Scenarios

Scenario 1: The Emergency Email Your boss needs you to decline a client request professionally. Practice writing prompts for different situations:

- Declining due to capacity
- Declining due to budget mismatch
- Declining due to scope misalignment

Scenario 2: The Learning Challenge You need to understand a new industry quickly. Create prompts for:

- Industry overview for beginners
- Key players and competitive landscape
- Major trends and challenges
- Relevant terminology explained simply

Scenario 3: The Creative Block You're stuck on a project. Develop prompts for:

- Brainstorming fresh angles
- Overcoming specific obstacles
- Finding inspiration from other fields
- Breaking complex projects into steps

The Practice Tracking System

Create a prompt journal with three columns:

1. **Original Prompt**: What you first tried

2. **Refined Prompt**: Your improved version

3. **Key Learning**: What made the difference

Example Entry:

- Original: "Marketing ideas"

- Refined: "Generate 5 unconventional marketing tactics for a B2B cybersecurity startup targeting banks. Budget: $5K/month. Focus on building trust and demonstrating expertise without being salesy."

- Learning: Specificity (B2B cybersecurity for banks) and constraints ($5K budget) yielded actionable ideas instead of generic suggestions

Common Patterns in Successful Prompts

Through these exercises, you'll notice patterns:

The Setup Pattern: "Context: [situation] Challenge: [specific problem] Request: [what you need] Constraints: [limitations] Format: [how you want it]"

The Progressive Detail Pattern: Start broad → Add context → Specify format → Include examples → Set constraints

The Comparison Pattern: "Compare [Option A] vs [Option B] for [specific use case]. Consider factors: [list 3-5 factors]. Present as: [desired format]."

Your Daily Practice Routine

Morning (5 minutes): Transform one weak prompt into a strong one **Midday (10 minutes)**: Practice role-based prompting for a real task **Evening (5 minutes)**: Reflect on the day's ChatGPT interactions and note improvements

The David Transformation

Remember David from the introduction? After two hours of exercises, he created this prompt for his work:

"Acting as a senior financial analyst specializing in tech valuations, help me build a DCF model framework for a Series B SaaS company. Revenue: $10M ARR, growing 100% YoY, 85% gross margins. Create:

1. Key assumptions checklist

2. 5-year projection template structure

3. Sensitivity analysis variables

4. Common pitfalls in SaaS valuations Format as structured guide I can follow in Excel."

His comment: "I never thought I'd be teaching ChatGPT prompting to my team, but here we are. The exercises made it click – it's not about perfect prompts, it's about thinking clearly and communicating precisely."

Your Practice Challenge

For the next week, commit to:

1. **Transform 3 weak prompts daily** using the exercises above

2. **Try one new constraint technique** each day

3. **Practice chain-of-thought prompting** for complex tasks

4. **Keep a prompt journal** of what works

5. **Share one success** with a colleague

The Muscle Memory Effect

Like learning any skill, prompt engineering develops through practice. These exercises aren't just about better ChatGPT outputs – they're about developing clarity of thought and communication.

Each time you transform a vague request into a precise prompt, you're strengthening neural pathways. Each iteration teaches you what works. Each constraint pushes creative solutions.

The goal isn't perfection – it's progress. Every prompt you improve, every iteration you attempt, every exercise you complete builds your capability.

So dive in. Make mistakes. Refine relentlessly. The pool's warm, and the only way to learn swimming is to get wet. Your prompting journey starts with the next exercise you try.

What prompt will you transform first?

Lesson 3.4: Troubleshooting Bad Responses – When ChatGPT Gets It Wrong

"This stupid AI is broken!"

Those were the first words I heard from Jennifer, a normally composed executive coach, when she called me in frustration. "I asked it to write a simple follow-up email to a client, and it gave me this robotic, corporate garbage that sounds nothing like me. I've tried five times. Same terrible result."

"Show me your prompt," I said.

"Write a follow-up email to my client."

I suppressed a smile. Jennifer wasn't dealing with a broken AI – she was experiencing the universal truth of ChatGPT: unclear instructions produce unclear results. Garbage in, garbage out. But here's the good news: every bad response is a diagnostic tool telling you exactly how to get better results.

The Diagnostic Mindset

When ChatGPT gives you a bad response, resist the urge to blame the AI. Instead, put on your detective hat. Bad responses are clues pointing to specific prompt problems.

Think of it like this: If you asked a human assistant to "write a follow-up email to my client" with zero context, what would they produce? Probably something generic and off-target. ChatGPT faces the same challenge – it can only work with what you give it.

The Five Categories of Bad Responses

Through thousands of troubleshooting sessions, I've identified five main types of problematic outputs, each with specific fixes:

1. The Generic Glob Symptoms: Vague, could-apply-to-anyone content full of buzzwords and clichés **Root Cause**: Insufficient context and specificity **Example**:

- Prompt: "Write about leadership"

- Result: "Leadership is about inspiring others... effective communication... blah blah..."

The Fix: Add context, audience, and unique angle

- Better: "Write 300 words about leadership challenges specific to remote engineering teams, focusing on asynchronous communication and maintaining technical standards. Audience: new engineering managers. Tone: practical, not preachy."

Real-World Fix: Jennifer's email problem was classic Generic Glob. Her refined prompt: "Write a warm follow-up email to Sarah, my executive coaching client. We just completed session 3 of 6, focused on delegation skills. Reference her breakthrough about trusting her team with the quarterly planning. Suggest scheduling our next session and ask how her delegation experiment went this week. Tone: supportive friend who's also professional. 150 words max."

Result: An email that sounded exactly like Jennifer.

2. The Wrong Tone Disaster Symptoms: Right content, completely wrong voice **Root Cause**: Missing or mismatched tone guidance **Example**:

- Prompt: "Explain our refund policy"
- Result: Cold, legal-sounding text that scares customers

The Fix: Explicitly define tone with examples or comparisons

- Better: "Explain our 30-day refund policy in a friendly, reassuring tone. Sound like a helpful customer service rep, not a legal document. Use 'you' and 'we' language. Include an example of how easy the process is."

Case Study: The Yoga Studio Maria's yoga studio needed website copy. First attempt produced "corporate wellness speak." Fixed prompt: "Write website copy for my neighborhood yoga studio. Tone: warm, inclusive, slightly spiritual but not intimidating. Like talking to a wise friend over tea, not a fitness influencer. Emphasize community and personal growth over physical achievement."

3. The Length Catastrophe Symptoms: Way too long or frustratingly short **Root Cause**: No length constraints specified **Example**:

- Prompt: "Summarize this article"
- Result: Either one sentence or three pages

The Fix: Always specify length

- Better: "Summarize this article in exactly 3 bullet points, each 1-2 sentences"
- Or: "Expand this concept into 500-600 words"
- Or: "Give me both a one-sentence summary and a one-paragraph explanation"

Pro Tip: ChatGPT understands various length measures:

- Word count (150 words)
- Character count (280 characters for tweets)
- Structural limits (5 bullet points, 3 paragraphs)
- Time estimates (2-minute read)

4. The Missed Point Special Symptoms: ChatGPT answers a different question than you asked **Root Cause**: Ambiguous phrasing or buried main point **Example**:

- Prompt: "I'm launching a new product next month and worried about competition and pricing and marketing, help"
- Result: Generic business advice missing your actual concern

The Fix: Lead with your main question

- Better: "Help me price my new productivity app. Context: launching next month, main competitor charges $9.99/month, my app has 3 unique features they don't. Suggest pricing strategy with reasoning."

Diagnostic Question: Can someone identify your main need within 5 seconds of reading your prompt? If not, rewrite.

5. The Hallucination Station Symptoms: Confident-sounding false information **Root Cause**: Asking for specific facts ChatGPT doesn't know **Example**:

- Prompt: "What were Apple's Q3 2024 revenues?"
- Result: Plausible but fictional numbers

The Fix: Avoid requests for specific facts, especially:

- Recent events or data
- Specific statistics without providing them
- Obscure technical details
- Real-time information

Better Approach:

- Provide the data: "Given that Apple's Q3 2024 revenues were $81.8B (you provide this), analyze..."
- Ask for frameworks: "Explain how to analyze quarterly revenue reports"
- Request general knowledge: "What factors typically impact tech company revenues?"

The Progressive Troubleshooting Method

When you get a bad response, follow this systematic approach:

Step 1: Diagnose the Category Which of the five problems are you facing? Generic? Wrong tone? Wrong length? Missed point? Hallucination?

Step 2: Apply Targeted Fix Use the specific remedy for that category.

Step 3: Iterate Intelligently Don't just say "try again." Give specific improvement instructions:

- "Good structure, but make the tone more conversational"
- "Right length, but focus more on benefits rather than features"
- "Better, but add specific examples for each point"

Step 4: Build on Success When you get a good section, keep it: "I like paragraphs 1 and 3. Rewrite paragraph 2 to match their style and add a stronger conclusion."

Real-World Troubleshooting Session

Watch how Mark, a real estate agent, troubleshot his way to success:

Attempt 1: "Write property description" **Result**: Generic house description **Diagnosis**: Generic Glob

Attempt 2: "Write description for 3-bed colonial in Westchester" **Result**: Better but still generic **Diagnosis**: Still missing unique details

Attempt 3: "Write 150-word description for $850K colonial in Scarsdale. Highlight: renovated kitchen (2023), original hardwood, walking distance to train, Blue Ribbon schools. Audience: NYC families looking to suburbs. Tone: sophisticated but warm." **Result**: Compelling description that helped sell the house

Attempt 4: "Perfect, but add a line about the private backyard perfect for entertaining" **Result**: Final version that Mark now uses as a template

The Advanced Troubleshooting Toolkit

Technique 1: The Comparison Method When output is wrong but you can't articulate why: "That's not quite right. Here's an example of the style I want: [paste good example]. Now rewrite your response to match this tone and approach."

Technique 2: The Exclusion Method When ChatGPT keeps including things you don't want: "Good attempt, but specifically avoid: corporate jargon, passive voice, and generic statements like 'in today's fast-paced world.' Focus on concrete, specific details."

Technique 3: The Building Block Method For complex tasks, troubleshoot in pieces: "Let's break this down. First, just give me an outline. [Troubleshoot outline] Good, now write just the introduction. [Troubleshoot intro] Perfect, now section 1..."

Technique 4: The Role Correction When the perspective is off: "You're writing too formally. Remember, you're a friendly neighborhood bakery owner talking to regular customers, not a corporate marketing department."

Common Troubleshooting Mistakes to Avoid

Mistake 1: The Rage Regenerate Hitting regenerate repeatedly without changing anything. Definition of insanity.

Mistake 2: The Vague Complaint "Make it better" or "That's not what I want" without specifics.

Mistake 3: The Kitchen Sink Addition Adding more and more requirements instead of fixing core issues.

Mistake 4: The Tone Whiplash "Make it professional but casual but authoritative but friendly but..."

Mistake 5: Fighting ChatGPT's Nature Trying to make it do things it fundamentally can't, like real-time data or perfect math.

Your Troubleshooting Cheat Sheet

Print this out and keep it handy:

Bad Response?

1. **Too Generic?** → Add context, specifics, unique angle

2. **Wrong Tone?** → Define voice explicitly with examples

3. **Wrong Length?** → Specify exact length requirements

4. **Missed Point?** → Lead with main question, clarify structure

5. **False Info?** → Provide data or ask for frameworks instead

Still Stuck?

- Start fresh in new conversation
- Break complex tasks into steps
- Provide examples of what you want
- Use process of elimination
- Ask ChatGPT to explain its interpretation

The Jennifer Resolution

Remember Jennifer from the introduction? After our troubleshooting session, she became a ChatGPT power user. "I realized bad responses weren't ChatGPT failing – they were ChatGPT teaching me to communicate better. Now when I get a bad response, I think 'What information did I forget to include?' It's made me a clearer thinker overall."

Her latest insight: "I keep a file of prompts that work perfectly for different situations. When something goes wrong, I compare the failed prompt to successful ones. The missing element is always obvious in hindsight."

Your Troubleshooting Practice

This week, deliberately collect bad responses. Instead of getting frustrated:

1. Diagnose which category of problem you're facing

2. Apply the specific fix

3. Document what worked

4. Build your personal troubleshooting playbook

Remember: Every bad response is a teacher. Every fix is a skill upgrade. Every troubleshooting session makes you a better prompter.

The goal isn't to avoid bad responses – it's to quickly diagnose and fix them. Master troubleshooting, and ChatGPT transforms from an occasionally frustrating tool to a consistently reliable partner.

What bad response will you fix first?

Lesson 3.5: Good vs. Bad Prompt Examples – Learning by Comparison

"I don't get it," said Robert, showing me his ChatGPT conversation. "My colleague Lisa and I asked for the same thing – help with our company newsletter. She got this amazing, engaging piece. I got corporate word soup. We work at the same company!"

I pulled up both their prompts:

Robert's prompt: "Write company newsletter"

Lisa's prompt: "Write the opening section (150 words) for our monthly internal newsletter at TechFlow (50-person B2B software company). Audience: mix of engineers, sales, and support staff. Tone: celebratory but not cheesy – we just landed our biggest client ever. Include: shoutout to sales team, teaser about upcoming product feature, reminder about Friday's summer BBQ. Avoid: corporate speak or fake enthusiasm. Write like you're a team member excited to share good news with colleagues."

The difference in their outputs wasn't luck – it was precision. Let me show you exactly what separates good prompts from bad ones through real-world comparisons.

Category 1: Writing Tasks

Bad Prompt: "Write a blog post about productivity"

Why It Fails:

- No target audience specified
- No length or structure guidance
- Generic topic without unique angle
- No tone or style direction

Result: Generic listicle about "5 Tips to Boost Productivity" full of advice like "eliminate distractions" and "take breaks."

Good Prompt: "Write a 600-word blog post about productivity specifically for emergency room nurses. Focus on managing energy during 12-hour shifts. Include: 3 practical strategies they can implement immediately, 1 brief story example, and acknowledgment of the unique challenges they face. Tone: empathetic colleague who gets it, not preachy outsider. Avoid: generic advice that ignores healthcare reality."

Why It Works:

- Crystal clear audience (ER nurses)
- Specific angle (energy during long shifts)
- Detailed structure requirements
- Appropriate tone defined

- Explicit exclusions

Result: Targeted, valuable content that resonates with the specific audience.

Category 2: Business Communication

Bad Prompt: "Help me with difficult conversation"

Result: Generic advice about active listening and staying calm.

Good Prompt: "I need to tell my top performer they didn't get the promotion they expected. Context: They're excellent but need 6 more months of leadership experience. I'm worried they'll quit. Help me plan this conversation with: 1) Opening that shows I value them, 2) Clear explanation without false promises, 3) Specific development plan, 4) Questions to gauge their reaction. Keep it conversational, not script-like."

Result: Actionable conversation framework tailored to the specific situation.

Real-World Impact: Manager Tom used the good prompt approach: "My employee not only stayed but thanked me for the clearest career conversation they'd ever had. The specific development plan ChatGPT helped create showed I was invested in their growth."

Category 3: Learning and Education

Bad Prompt: "Explain quantum computing"

Why It Fails:

- No audience level specified

- No context for why they're learning
- No format or length guidance
- Too broad without focus

Result: Technical explanation that's either too basic or too complex.

Good Prompt: "Explain quantum computing to a CEO who needs to decide whether to invest in quantum technology for their financial services firm. Cover: 1) Basic concept using banking analogy, 2) Potential applications in finance within 5 years, 3) Current limitations, 4) Investment considerations. 300 words, executive-friendly language, focus on business impact not technical details."

Why It Works:

- Clear audience and their need
- Specific areas to cover
- Appropriate analogies suggested
- Length and style specified
- Business focus over technical

Result: Executive briefing that enables informed decision-making.

Category 4: Creative Projects

Bad Prompt: "I need creative ideas"

Result: Random list of unrelated creative project ideas.

Good Prompt: "Generate 5 creative marketing campaign ideas for our eco-friendly water bottle brand targeting college students.

Constraints: $5K budget, must be social media friendly, should highlight our lifetime warranty and plastic-free mission. For each idea: catchy name, basic concept, why it would resonate with Gen Z, and estimated reach."

Result: Targeted, actionable campaign ideas with implementation details.

Success Story: Sarah's startup used this approach: "ChatGPT suggested a 'Plastic Funeral' campaign where students ceremoniously 'bury' their last disposable bottle. It went viral on TikTok. Would never have thought of that with a vague prompt."

Category 5: Problem-Solving

Bad Prompt: "How do I get more customers?"

Why It Fails:

- No business context
- No current situation analysis
- No constraints or resources mentioned
- Too broad to be actionable

Result: Generic marketing advice applicable to any business.

Good Prompt: "My neighborhood coffee shop has been open 6 months. Current: 50 daily customers, average $8 spend, mostly morning rush. Goal: reach 100 daily customers in 3 months. Constraints: $1000 marketing budget, can't expand hours, limited to 2 staff. Neighborhood: young professionals and families. What are 5 specific, low-cost tactics to double customers? For each, explain implementation and expected impact."

Why It Works:

- Complete context provided
- Specific, measurable goal
- Clear constraints listed
- Relevant details included
- Actionable output requested

Result: Practical growth strategies tailored to the specific business situation.

Category 6: Technical Assistance

Bad Prompt: "Help with Excel"

Result: Basic Excel tutorial or random tips.

Good Prompt: "I have an Excel spreadsheet with 10,000 rows of sales data. Columns: Date, Product, Sales Rep, Amount, Region. I need to: 1) Find top 5 performing products by total revenue, 2) Compare sales rep performance by region, 3) Identify seasonal trends. Walk me through the specific formulas and pivot table setup. Assume intermediate Excel knowledge."

Result: Step-by-step instructions for exactly what you need to accomplish.

The Prompt Transformation Exercise

Let's transform more bad prompts into good ones:

Bad: "Write about customer service" **Good**: "Create a 1-page customer service recovery framework for retail associates dealing with angry customers. Include: 4-step de-escalation process, example phrases for common situations, and when to involve a manager. Format: easy-reference guide they can keep at the register."

Bad: "Marketing advice for my business" **Good**: "I run a dog grooming business in suburban Denver. 5 years old, 60% capacity, customers love us but we need more. Create 90-day marketing plan focused on filling our slow Tuesday/Wednesday slots. Budget: $500/month. Must leverage our strength: amazing before/after photos."

Bad: "Help me with time management" **Good**: "I'm a freelance graphic designer juggling 5 clients. Main problem: context switching kills my creativity. Design a daily schedule template that batches similar work, protects creative time, and includes client communication windows. Consider: I'm most creative ini the mornings, calls drain me, deadlines are non-negotiable."

The Pattern Recognition Guide

Through these examples, notice these patterns:

Bad Prompts Share:

- Vague objectives
- Missing context
- No constraints
- Generic requests
- Unclear desired output

Good Prompts Share:

- Specific situations

- Clear constraints

- Defined audience

- Structured output

- Unique angles

Your Prompt Upgrade Checklist

Before sending any prompt, check:

□ **WHO**: Is the audience/role clear? □ **WHAT**: Is the specific output defined? □ **WHY**: Is the context/purpose included? □ **HOW**: Are format/style/tone specified? □ **CONSTRAINTS**: Are limitations clearly stated? □ **UNIQUE**: Is there a specific angle or focus?

Real-World Prompt Library

Here are tested prompts that consistently deliver:

For Email Writing: "Write a [length] email to [specific recipient] about [specific topic]. Context: [relevant background]. Tone: [specific description]. Include: [key points]. Avoid: [what not to include]."

For Analysis: "Analyze [specific thing] from the perspective of [specific role]. Consider factors: [list 3-5]. Present findings as: [format]. Focus on: [particular aspect]. Length: [word count]."

For Creative Tasks: "Generate [number] ideas for [specific project] with constraints: [list limitations]. Target audience: [description]. For each idea provide: [required elements]. Evaluation criteria: [what makes a good option]."

The Robert and Lisa Epilogue

Remember Robert from the introduction? After seeing the comparison, he transformed his prompting approach. "I thought being brief was being respectful of AI's time," he laughed. "I didn't realize I was making its job harder."

His latest newsletter prompt: "Write opening paragraph (100 words) for TechFlow's December newsletter. Tone: warm but professional, like a year-end letter from a friend. Include: gratitude for team's hard work, excitement about January's product launch (AI assistant for sales teams), and invitation to holiday party (Dec 15, ugly sweater theme). Avoid: corporate clichés or forced holiday cheer. Write from the perspective of a CEO who genuinely appreciates the team."

The result? "Best newsletter opening we've ever had. Three people told me it actually made them smile."

Your Comparison Practice

This week:

1. **Collect your "bad" prompts** that gave weak results
2. **Analyze why they failed** using the patterns above
3. **Rewrite using the good prompt structure**
4. **Compare the outputs** side by side
5. **Document the differences** for future reference

The Ultimate Insight

The difference between good and bad prompts isn't complexity – it's clarity. Good prompts aren't necessarily longer; they're more precise. They don't require special formatting or magic words; they require clear thinking about what you actually want.

Lisa put it best: "Once I started writing prompts like I was briefing a talented but literal assistant, everything clicked. The key is remembering that ChatGPT can't read between the lines – so don't leave anything between them."

Master this distinction, and you'll never wonder why your outputs are hit-or-miss again. Every prompt becomes an opportunity to get exactly what you need – if you're willing to clearly ask for it.

CHAPTER 4

Everyday Uses (Your Personal AI Helper)

~

Opening Story: The Day That Changed Everything

Julia Martinez sat at her kitchen table at 5:47 AM, laptop open, coffee getting cold, surrounded by the chaos of modern life. As a single mom running a small catering business while finishing her degree, every minute mattered. Yet here she was, stuck on three simple tasks: writing a thank-you note to a client, helping her daughter with a science project about volcanoes, and figuring out a week's worth of healthy dinners on a tight budget.

"I just need more hours in the day," she muttered, then remembered her sister's advice about ChatGPT. Skeptically, she opened it and typed: "I need help writing a professional thank you note to a wedding client."

Fifteen minutes later, Julia had.

- A heartfelt, professional thank-you note that sounded exactly like her
- A kid-friendly explanation of volcanoes with a simple experiment using baking soda
- A complete meal plan with shopping list, all under $100

"It was like having a personal assistant, tutor, and meal planner show up at my kitchen table," Julia told me later. "But the real magic? It gave me back two hours that morning. Two hours I spent actually growing my business instead of struggling with routine tasks."

Julia's story isn't unique. Millions are discovering that ChatGPT isn't just for tech experts or big corporations – it's a practical tool for the overwhelming demands of daily life.

Lesson 4.1: Writing Help – Your AI Wordsmith

We all write more than ever – emails, texts, posts, letters, lists. Yet for many, putting thoughts into words remains a daily struggle. ChatGPT transforms this struggle into a strength, but only if you know how to collaborate with it effectively.

The Writing Partnership Mindset

Think of ChatGPT not as a replacement writer but as a writing partner who:

- Never judges your first draft
- Offers unlimited revisions
- Knows every writing style
- Works at your pace
- Available 24/7

The key word is "partner." You bring the ideas, context, and authenticity. ChatGPT brings structure, polish, and possibilities.

Email Mastery: From Dreaded to Done

Email consumes an average of 28% of the work week. Let's reclaim that time.

The Thank-You Note Transformation Julia's original attempt: Staring at blank screen, typing and deleting "Thank you for choosing us" five times.

Her ChatGPT prompt: "Help me write a warm thank-you email to the Chen family who hired my catering company for their daughter's wedding last Saturday. Mention: the bride loved our signature lavender lemonade, we were honored to be part of their special day, and we'd love to cater their future celebrations. Tone: professional but heartfelt, like writing to friends. 150 words."

The result: A note so genuine the Chens referred three new clients.

Email Templates That Actually Work

The Polite No Prompt: "Write a polite email declining a meeting invitation. Reason: fully booked this week. Offer: to review their materials and provide feedback via email instead. Tone: helpful, not dismissive. 100 words."

The Follow-Up That Gets Responses Prompt: "Write a friendly follow-up email. Context: sent proposal 1 week ago, no response. Don't be pushy but convey gentle urgency. Mention: happy to answer questions or adjust proposal. Include clear call-to-action. 125 words."

The Difficult Conversation Starter Prompt: "Help me write an email to my landlord about needed repairs. Issues: leaking bathroom faucet (3 weeks), broken closet door. Tone: firm but respectful. Include: impact on daily life, request for timeline. Avoid: threatening or aggressive language. 150 words."

Beyond Email: Writing for Life

The Social Media Struggle Mark runs a local bike shop but freezes when facing Instagram. "I know bikes, not hashtags," he complained.

His solution: "Write 5 Instagram captions for my bike shop. Posts about: Monday motivation, new bike day, maintenance tips, local trail recommendation, weekend group ride. Include relevant hashtags. Voice: enthusiastic local bike expert, not salesy corporation. Each 50-75 words."

Result: Engagement up 200% in one month.

The Personal Writing Assistant

Condolence Messages "Help me write a sympathy card message for my coworker whose mother passed away. We're friendly but not close friends. Express genuine sympathy without being overly personal. 3-4 sentences."

Complaint Letters That Work "Write a complaint letter to airline about lost luggage. Flight: UA123, May 15. Impact: missed business presentation. Tone: professional, not angry. Goal: fair compensation and luggage return. Include: next steps expected. 250 words."

Dating Profile Magic "Help me write a dating profile bio. About me: teacher, love hiking and cooking, have a golden retriever named Max. Looking for: genuine connection, someone who laughs at dad jokes. Tone: warm and approachable, hint of humor. Avoid: clichés like 'love to laugh.' 150 words."

The Revision Revolution

First drafts don't have to be perfect anymore. Watch this progression:

Round 1: "Write a cover letter for marketing manager position" Result: Generic

Round 2: "Make it specific to digital marketing role at sustainable fashion brand. Highlight my social media success and passion for eco-friendly business." Result: Better but stiff

Round 3: "Perfect content. Now make it sound more conversational and enthusiastic. Show personality while staying professional." Result: Interview landed

Writing Productivity Hacks

The Brain Dump Method

1. Dump all your thoughts to ChatGPT, however messy
2. Ask it to organize into coherent structure
3. Refine the organization
4. Polish the language
5. Add your personal touches

The Template Library Build your personal collection:

- Email templates for common situations
- Social media post structures
- Report formats
- Letter frameworks

Save prompts that generate your perfect tone.

The Voice Cloning Technique "Here's an email I wrote that perfectly captures my voice: [paste example]. Now write a similar email about [new topic] matching this tone and style."

Real-World Writing Wins

The Grant Writer's Secret Nonprofit director Sharon: "I used to spend days on grant applications. Now I provide ChatGPT with our mission, specific program details, and grant requirements. It creates the structure and formal language. I add our stories and data. Time saved: 70%. Grants won: up 40%."

The Newsletter Transformation Restaurant owner David: "Weekly newsletters went from 3-hour nightmare to 30-minute task. I tell ChatGPT our specials, events, and one fun story. It creates engaging newsletter. I tweak and send. Customers say they actually look forward to them now."

The Academic Advantage Student Emma: "I don't use ChatGPT to write my papers. I use it to organize my thoughts, improve my thesis statements, and make my writing clearer. My professor noticed the improvement and asked what changed. I said I got better at editing – which is true."

Common Writing Pitfalls to Avoid

Don't Lose Your Voice ChatGPT can write in any style, but it shouldn't replace yours. Use it to enhance, not erase, your unique voice.

Don't Skip the Personal Touch Always add specific details only you know. ChatGPT provides structure; you provide soul.

Don't Forget to Fact-Check Especially for professional writing, verify any facts or figures ChatGPT includes.

Don't Use Without Reading Never send ChatGPT output without reading and personalizing. Your reputation is worth the extra minute.

Your Daily Writing Workflow

Morning Email Batch

- Draft all responses with ChatGPT
- Personalize each
- Send in one focused session

Content Creation Hour

- Generate week's social media posts
- Create blog post outlines
- Draft marketing materials

End-of-Day Wrap

- Thank you notes
- Follow-ups
- Tomorrow's important emails

The Writing Confidence Effect

Here's what Julia discovered: "ChatGPT didn't just help me write better – it helped me become a better writer. By seeing how it structured thoughts and chose words, I learned. Now I often write first drafts myself, then use ChatGPT for polish. It's like having a writing coach who never gets tired of helping."

Your Writing Action Plan

This week, use ChatGPT to:

1. **Clear your email backlog** - Those messages you've been avoiding

2. **Create templates** for your five most common writing tasks

3. **Write something personal** you've been postponing

4. **Improve existing content** - Old bio, about page, or résumé

5. **Start that project** - Blog, newsletter, or book outline

The Bottom Line

Writing is thinking made visible. ChatGPT doesn't think for you – it helps you express your thoughts more clearly, professionally, and efficiently. It's the difference between dreading the blank page and approaching it with confidence.

Master ChatGPT for writing, and you don't just save time. You communicate better, connect more deeply, and express yourself more fully. In a world built on written communication, that's a superpower worth developing.

What will you write first?

Lesson 4.2: Learning & Understanding – Your AI Tutor

Professor Chen had been teaching calculus for 20 years, but his daughter's question stumped him: "Dad, I get that derivatives measure change, but WHY do we flip the fraction when we integrate?"

He started explaining using limits and Riemann sums, watching her eyes glaze over. Then he remembered ChatGPT. "Explain why we flip fractions in integration to a visual learner in high school. Use an analogy they can picture."

ChatGPT's response used a brilliant analogy about filling a swimming pool versus measuring flow rate. His daughter's eyes lit up. "Oh! So integration is like going from gallons-per-minute to total gallons, and that's why the units flip!"

"In 20 years of teaching," Professor Chen told me, "I never thought to explain it that way. ChatGPT didn't just help my daughter – it made me a better teacher."

The Learning Revolution in Your Pocket

ChatGPT represents something unprecedented in human history: a patient, knowledgeable tutor available to anyone, anytime, for free. It never gets frustrated, never judges your questions, and can explain the same concept 50 different ways until one clicks.

But here's the key: It's not about ChatGPT knowing everything – it's about ChatGPT helping you understand anything.

The Five-Layer Learning Method

Through working with hundreds of learners, I've developed the Five-Layer Method for using ChatGPT to truly understand complex topics:

Layer 1: The ELI5 (Explain Like I'm 5) Start with the simplest possible explanation.

Example: "Explain cryptocurrency like I'm 5 years old" Result: "Imagine special digital coins that live in computers. Instead of a bank keeping track, lots of computers work together to remember who owns which coins."

Layer 2: The Real-World Connection Connect to something familiar.

"Now explain cryptocurrency using a school cafeteria analogy" Result: "Think of lunch money cards that students trade directly without the cafeteria lady. Everyone has a notebook to track trades, and they all check each other's math."

Layer 3: The Technical Foundation Add appropriate complexity.

"Now explain the actual technology behind cryptocurrency for a beginner" Result: Blockchain explanation with proper terminology but accessible language.

Layer 4: The Practical Application Make it relevant to your life.

"How might cryptocurrency affect my small business in the next 5 years?" Result: Specific scenarios and considerations for business owners.

Layer 5: The Deep Dive Explore nuances and edge cases.

"What are the main criticisms of cryptocurrency and are they valid?" Result: Balanced analysis of environmental, regulatory, and practical concerns.

Subject-Specific Learning Strategies

For Science and Math The key is multiple representations.

Poor approach: "Explain photosynthesis" Better approach: "Explain photosynthesis three ways: 1) As a simple story, 2) With a cooking analogy, 3) With the actual chemical process. Then show me how to remember it."

Real Success: High school student Marcus was failing chemistry. His ChatGPT strategy: "After each class, I explain what I learned to ChatGPT and ask it to correct my understanding. Then I have it create practice problems. My grade went from D to B+ in one semester."

For History and Social Studies Context and connections matter most.

Instead of: "Tell me about World War II" Try: "Help me understand the causes of WWII by connecting them to issues in 1930s daily life. What would an average person have noticed changing?"

Teacher Testimony: "I have students use ChatGPT to create 'history dialogues' – conversations between historical figures. They have to research enough to make the dialogue accurate. It's transformed how they engage with history." - Ms. Rodriguez, AP History

For Languages ChatGPT becomes your conversation partner.

Spanish learner Maria's routine:

1. "Let's have a simple conversation in Spanish about my day"

2. "Correct my Spanish mistakes from above and explain why"

3. "Teach me 5 new phrases related to what we discussed"

4. "Create a short story using those phrases"

Result: "More Spanish practice than I ever got in class, and it's actually relevant to my life."

For Professional Skills Learn by doing with guidance.

Example: "I want to learn Excel pivot tables. Walk me through creating one for this sales data: [paste data]. Explain each step and why we do it. Then give me a practice problem."

The Personalized Learning Advantage

Learning Style Adaptation Visual learner? "Explain using diagrams I can picture" Auditory learner? "Explain using rhythm or music analogies" Kinesthetic learner? "Explain using physical movement analogies"

Pace Control "Wait, explain that last part again but slower" "I get it, move to the next concept" "Give me a harder example now"

Interest Integration Basketball fan learning physics? "Explain momentum using basketball examples" Gamer learning history? "Explain the Roman Empire like it's a strategy game"

Common Learning Challenges and Solutions

Challenge: Information Overload Solution: "Explain [topic] in exactly 3 key points. Then we'll dive deeper into each."

Challenge: Missing Prerequisites Solution: "I'm trying to understand [advanced topic]. What basics do I need to know first? Create a learning path."

Challenge: Retention Issues Solution: "Create a memorable mnemonic/acronym/story to help me remember [concept]"

Challenge: Application Gap Solution: "I understand the theory of [concept]. Give me 5 real-world scenarios where I'd use this."

The Study Session Revolution

Here's how successful students structure ChatGPT study sessions:

Pre-Class Prep (10 minutes) "I'm about to learn about [topic] in class. What are the 3-5 key concepts I should watch for? What typically confuses students?"

Post-Class Review (15 minutes) "Here are my notes from today's class on [topic]. Help me identify what I understood correctly and what needs clarification."

Homework Helper (As needed) "I'm stuck on this problem: [problem]. Don't solve it for me, but give me a hint about which concept to apply."

Test Preparation (30 minutes) "Create a practice test on [topics] at [difficulty level]. After I answer, explain why each answer is right or wrong."

Real-World Learning Victories

The Career Changer Tom, 42, transitioning from retail to tech: "I used ChatGPT to learn programming. Not to write code for me, but to explain concepts, debug my attempts, and suggest projects. Six months later, I landed a junior developer job."

The Curious Retiree Betty, 68: "I always wanted to understand investing but felt too old to start. ChatGPT explains things patiently, remembers what confuses me, and never makes me feel stupid for asking. I now manage my own portfolio."

The Struggling Student Jamie, dyslexic high schooler: "ChatGPT explains things in ways that work for my brain. I can ask it to use more visual descriptions or break things into smaller chunks. My grades improved, but more importantly, I actually enjoy learning now."

The Metacognition Bonus

The hidden benefit: ChatGPT teaches you how to learn.

"I noticed I understand better when concepts connect to things I already know," shared student Alex. "Now I always ask for analogies. ChatGPT taught me how my brain works best."

Your Learning Action Plan

Week 1: Explore One Curiosity Pick something you've always wanted to understand. Use the Five-Layer Method.

Week 2: Tackle a Practical Skill Choose a skill relevant to work or life. Create a structured learning plan with ChatGPT.

Week 3: Revisit Old Struggles That subject you "just couldn't get"? Try again with ChatGPT as your tutor.

Week 4: Teach to Learn Explain something you know to ChatGPT and ask it to identify gaps in your explanation.

The Learning Mindset Shift

Professor Chen's insight captures it perfectly: "ChatGPT doesn't replace teachers – it empowers learners. It's the difference between passive education and active understanding. My students who use it don't just get better grades; they become better thinkers."

Learning Best Practices

1. **Always verify important facts** - ChatGPT explains well but can make errors

2. **Use it for understanding, not shortcuts** - Learn the concept, don't just copy answers

3. **Practice active learning** - Explain back to ChatGPT to test understanding

4. **Create connections** - Always ask how new knowledge relates to what you know

5. **Embrace confusion** - "I don't understand X about this" leads to breakthrough moments

The Future of Personal Education

We're entering an era where anyone can learn anything. Not because information is available – Google did that. But because explanation, adaptation, and patient guidance are available infinitely, personalized to exactly how your brain works.

ChatGPT isn't just an AI tutor. It's a learning amplifier that adapts to you, works at your pace, and never gives up on helping you understand. Master this tool, and you master the ability to master anything.

What will you finally understand today?

Lesson 4.3: Planning & Organizing – Your AI Assistant

"I'm drowning in chaos," Michelle confessed, showing me her desk covered in sticky notes, three different calendars, and a to-do list that seemed to reproduce overnight. As a marketing manager and mother of twins, she juggled approximately 847 things daily.

"Watch this," I said, opening ChatGPT. "Let's organize your entire week in 15 minutes."

Michelle laughed. "In 15 minutes, I usually just decide which fire to put out first."

Twenty minutes later (okay, I was slightly off), Michelle had:

- A prioritized weekly schedule with time blocks
- Meal plans with prep shortcuts
- A project timeline with delegated tasks
- A family activity calendar
- A system for maintaining it all

"It's like having a personal assistant who actually understands my life," she marveled. "Why didn't anyone tell me ChatGPT could do THIS?"

The Organization Multiplier Effect

Most people use planners. Smart people use systems. But the smartest people use AI to create personalized systems that actually stick. ChatGPT doesn't just help you plan – it helps you design planning approaches that work with your brain, not against it.

Planning That Actually Happens

The Weekly Architecture Method Instead of random scheduling, build your week like an architect:

Prompt: "Help me design my ideal week. Context: Marketing manager, 2 kids (age 7), side consulting business. Priorities: Hit work deadlines, family dinner 4x/week, exercise 3x, advance side business. Constraints: Kids' soccer Tuesday/Thursday 6pm, weekly team meeting Wednesday 10am. Create time-blocked schedule that's realistic, not aspirational."

ChatGPT creates a framework considering:

- Energy levels throughout the day
- Transition time between activities
- Buffer zones for the unexpected
- Protected time for priorities

The Reality Check Revision Michelle's insight: "The first draft was too ambitious. So I said: 'This assumes I'm a robot. Add 15-minute buffers between tasks, 30 minutes daily for unexpected issues, and make Friday afternoon lighter for week wrap-up.' The revised version actually worked."

Meal Planning Without the Mental Load

The Constraint-Based Meal Plan "Create 5-day dinner plan. Constraints: Budget $100, one vegetarian kid, max 30-minute cook time, use instant pot when possible, include leftovers for lunch. Family favorites: tacos, pasta, stir-fry. Avoid: mushrooms, spicy food. Include shopping list organized by store sections."

Result: Monday's instant pot chili becomes Tuesday's loaded baked potatoes. Wednesday's chicken becomes Thursday's chicken quesadillas. Efficient, economical, everyone's happy.

The Prep Strategy Session "Based on this meal plan, create a Sunday prep strategy that takes 1 hour and makes weeknight cooking faster."

ChatGPT identifies:

- What to chop in advance
- Which sauces to pre-make
- What to marinate ahead
- Efficient prep order

Project Management for Real Life

The Backwards Planning Method "I need to plan my daughter's birthday party for June 15. Work backwards to create a timeline with all tasks, when to do them, and what can be delegated. Include: 20 kids, park pavilion, unicorn theme, budget $300."

ChatGPT creates:

- 6 weeks out: Book pavilion, create guest list
- 4 weeks out: Send invitations, order decorations
- 2 weeks out: Confirm headcount, order cake
- 1 week out: Buy supplies, prep party favors
- Day before: All prep tasks listed

The Decision Tree Approach For complex planning: "I'm planning a vacation for family of 4. Budget: $3000. Create a decision tree: If we fly, then X. If we drive, then Y. Show how each choice affects other options."

The Daily Operations Manual

Morning Routine Optimizer "Design morning routine to get me and 2 kids out door by 7:30am. Currently it takes 90 minutes and is stressful. Include: breakfast, packing lunches, getting dressed. Find efficiency."

ChatGPT suggests:

- Clothes laid out night before
- Lunch components prepped Sunday
- Breakfast rotation (no daily decisions)
- Parallel processing (kids dress while you make lunch)
- Built-in buffer time

The Evening Reset Protocol "Create a 20-minute evening routine that sets up tomorrow's success. Include: kitchen clean, prep for morning, kid's school stuff, and one self-care element."

Event Planning Made Simple

The Complete Event Framework Prompt: "I'm organizing a client appreciation event. 50 people, afternoon networking, budget $2000. Create a complete plan including: timeline, vendor checklist, budget breakdown, contingency plans, and follow-up strategy."

Real example: Consultant Rachel used this approach: "ChatGPT thought of details I always miss – dietary restriction planning, parking instructions, name tag system. The event was my smoothest ever."

The Optimization Iterations

Version 1: Basic plan "Create study schedule for my certification exam in 6 weeks"

Version 2: Personalized reality "I can only study 8-10pm weekdays and Saturday mornings. Adjust plan accordingly"

Version 3: Life-proof planning "Add catch-up sessions for when life happens. Build in review days. Account for my weakness in technical sections"

Travel Planning Without Stress

The Comprehensive Approach "Plan 4-day trip to Seattle with teens. Interests: music, food, outdoors. Create an itinerary with: must-sees, restaurant options near each activity, rainy day alternatives, free time built in. Include estimated costs and transportation notes."

ChatGPT provides:

- Day-by-day itinerary with flexibility
- Restaurant options for picky eaters
- Teen-approved activities
- Budget tracking template
- Packing list based on activities

The Habit Installation System

The Micro-Step Method "I want to start exercising but always fail. Create a 30-day habit installation plan starting with stupidly small steps that gradually increase. Account for my tendency to go too hard then quit."

Week 1: 5-minute walk daily Week 2: 10-minute walk Week 3: Add 5 minutes strength Week 4: Full 20-minute routine

"The gradual approach worked where my usual 'gym every day starting tomorrow' always failed." - Michelle

Real-World Planning Victories

The Overwhelmed Entrepreneur "I used ChatGPT to plan my product launch. It created a Gantt chart in text, identified dependencies I missed, and suggested a promotional timeline. The launch was the smoothest ever. Revenue up 40%." - Jake, SaaS founder

The Busy Parent "ChatGPT helped me plan summer for 3 kids with different interests and no camps. Created weekly themes, daily activities, and budget tracker. First summer I wasn't counting days until school." - Parent of three

The Career Transitioner "Planned my transition from corporate to freelance over 6 months. ChatGPT helped create financial runway, client pipeline, and milestone markers. Quit my job exactly on schedule with 3 clients lined up." - Former corporate manager

Common Planning Pitfalls and Fixes

Pitfall: Over-optimistic scheduling **Fix**: "Add 20% more time to each task and include transition buffers"

Pitfall: Ignoring energy levels **Fix**: "Adjust schedule based on my energy: creative work in morning, meetings after lunch, admin when tired"

Pitfall: No flexibility **Fix**: "Include 'flex blocks' for unexpected tasks and opportunities"

Pitfall: Planning in isolation **Fix**: "Consider other people's schedules and needs in my planning"

Your Planning Transformation Toolkit

The Weekly Planning Session Every Sunday, 30 minutes:

1. Brain dump everything to ChatGPT
2. Have it organized by priority and urgency
3. Create realistic weekly schedule
4. Identify potential conflicts
5. Build in adjustment space

The Monthly Strategy Check First of month:

- Review what worked/didn't last month
- Adjust systems based on reality
- Plan major events and deadlines
- Create month-at-a-glance overview

The Quarterly Life Design Every three months:

- Assess life balance across areas
- Adjust priorities if needed
- Plan major projects or changes
- Create systems for new goals

The Michelle Transformation

Six months later, Michelle runs her department, her side business is thriving, and she actually makes it to her kids' soccer games. "The key wasn't doing more – it was planning smarter. ChatGPT helped me see my time differently. Instead of 24 hours of chaos, I have blocks of possibility."

Her best insight: "I stopped planning perfection and started planning for reality. ChatGPT helps me create systems that assume I'm human, life is messy, and flexibility beats rigidity every time."

Your Planning Action Challenge

This week:

1. **Map your current chaos** - Dump everything you're juggling into ChatGPT

2. **Create one system** - Pick your biggest pain point and design a solution

3. **Test and adjust** - Try the system for 3 days, then refine

4. **Build gradually** - Add one new planning element each week

5. **Share success** - Systems work better when others are involved

The Planning Paradigm Shift

ChatGPT doesn't just help you plan – it helps you become a planner. The difference? Planners don't just organize tasks. They design lives. They see patterns. They create systems that persist even when motivation doesn't.

Master planning with ChatGPT, and you master the architecture of your days. You stop reacting and start creating. You move from surviving to designing.

Your life is waiting to be organized. What will you plan first?

Lesson 4.4: Fun & Creativity – Your AI Muse

"I haven't written a poem since high school English class," laughed Robert, the 45-year-old accountant sitting across from me. "But my daughter's 16th birthday is tomorrow, and I want to give her something more meaningful than another gift card."

"What makes her special?" I asked, opening ChatGPT.

"She's fearless. Plays three sports, teaches herself guitar from YouTube, volunteers at the animal shelter. Always pushing boundaries. I'm... well, I'm an accountant."

Twenty minutes later, Robert was wiping his eyes as he read the poem ChatGPT helped him craft – not because the AI wrote beautiful words, but because it helped him express feelings he'd carried for 16 years but never knew how to articulate.

"It's still my love, my words, my memories," he said. "ChatGPT just helped me arrange them into something beautiful."

Creativity Isn't Just for "Creative People"

Here's the truth: Everyone has creative impulses. We just bury them under "I'm not creative" or "I don't have time" or "That's not practical." ChatGPT doesn't make you creative – it removes the barriers between your creative impulses and their expression.

The Poetry of Personal Moments

Birthday Poem Magic Robert's prompt evolution shows how ChatGPT helps refine personal creativity:

Attempt 1: "Write a poem for my daughter's 16th birthday" Result: Generic rhyming verses

Attempt 2: "Write a poem for Emma's 16th birthday. Include: her fearless spirit, how she taught herself guitar, her kindness to animals, how proud I am. From her dad who's better with numbers than words." Result: Better, but still not quite right

Attempt 3: "Perfect structure, but make it sound like a dad who chokes up at every milestone. Include the memory of teaching her to ride a bike and realizing she didn't need me to hold on anymore. Mix humor about my accounting brain with deep love." Result: The poem that made Emma cry (happy tears) and ask for it framed

Wedding Toast Transformation Best man Steve panicked: "I'm supposed to be funny but heartfelt. I'm neither."

His ChatGPT collaboration: "Help me write a best man toast. Context: Known groom since college (20 years), we started a failed food truck together, he met the bride at salsa dancing (I dragged him there), she made him eat vegetables and wake up before noon. Include: embarrassing story about the food truck catching fire, how she transformed him, my genuine happiness. Tone: 70% humor, 30% heartfelt. Length: 3 minutes."

Result: Standing ovation, requests for copies, bride's mother crying.

The Story Generator

Bedtime Story Revolution Parent trick: "Create a bedtime story where [child's name] is the hero who saves the day using [their favorite thing]. Include their stuffed animals as sidekicks and a lesson about [value you're teaching]."

Sarah's example: "Write a story where Luna (age 5) uses her love of painting to save a colorless world. Her stuffed bear Mr. Buttons and elephant Ellie help. Teach about sharing creativity with others. 5-minute read."

Family History Preservation "Turn this rough outline of my grandmother's immigration story into a narrative my kids will want to read. Include: leaving Italy at 17, working in a NYC garment factory, meeting grandpa at dance hall, opening bakery. Make it feel like an adventure, not a history lesson."

The Game and Activity Creator

Personalized Party Games "Create a scavenger hunt for 8-year-old's dinosaur birthday party. 10 kids, backyard setting, mix of active and thinking challenges. Include dinosaur facts they'll think are cool. End with finding the 'fossil' cake."

Rainy Day Savior "I have 3 bored kids (ages 5, 8, 10), basic craft supplies, and 2 hours to fill. Create 3 connected activities that build on each other and end with something they can show dad when he gets home."

ChatGPT created: Cardboard city → Story about city residents → Play presenting story with crafted puppets.

The Creative Problem Solver

Theme Party Genius "Planning an adult Halloween party but tired of typical themes. Suggest 5 unique themes with costume ideas that are creative but achievable, decoration suggestions using mostly items from dollar stores, and themed cocktail names."

Winning theme: "Extinct Things Party" - guests as discontinued products, old technology, extinct animals. Cocktail: "The Dodo" (because you'll be extinct if you have too many).

Gift Idea Generator "Need a gift for a person who has everything. Info: 60-year-old uncle, loves golf and cooking, just retired, budget $100. Suggest gifts that are thoughtful, not generic."

Best suggestion: Custom spice blend set with golf-themed names ("Birdie BBQ Rub," "Eagle Eye Steak Seasoning") plus handwritten recipe cards for each.

The Humor Helper

The Dad Joke Database "My kids groan at my jokes. Generate 10 actually funny dad jokes related to [current situation/holiday/event]."

Father's Day winner: "Why did the scarecrow win Father of the Year? He was outstanding in his field, just like me!"

The Roast Assistant "Help me write a funny but loving roast for my best friend's 40th birthday. Include: his obsession with lawn care, how he still dresses like 1995, his secret Taylor Swift fandom. Keep it playful, not mean."

Creative Projects That Stuck

The Recipe Reimaginer "Take my grandmother's boring-sounding 'Tuna Surprise Casserole' recipe and rewrite it like it's a trendy restaurant dish. Keep the ingredients but make it sound appealing to my foodie friends."

Result: "Vintage Atlantic Tuna Gratin with Artisanal Pasta, Garden Peas, and Golden Breadcrumb Crust" - Same dish, friends actually tried it.

The Memory Book Maker "Help me create a retirement memory book for my boss. I have 20 coworker quotes/stories. Organize them into themes, suggest creative section titles, and write transitions that flow. Add humor about his coffee addiction and fear of new technology."

The Confidence Creator

Karaoke Courage "I finally agreed to karaoke. Suggest 5 songs for someone who can't really sing but wants to have fun. Consider: male voices, crowd pleasers, songs where enthusiasm matters more than pitch."

ChatGPT's winning suggestion: "Sweet Caroline" - everyone sings the "BAH BAH BAH!"

The Love Letter Liberation "Help me write a love letter to my wife for our anniversary. We've been married for 12 years, have 3 kids, and barely have time for romance. I want to remind her she's still the girl I fell for, not just 'mom.' Include: how she dances in the kitchen, her terrible singing I love, how she makes ordinary moments special."

Result: "She cried. Good tears. Then danced in the kitchen. Still can't sing." - Happy husband

Your Creative Catalyst Toolkit

The Daily Creative Vitamin Each day, try one:

- Monday: Write a haiku about your morning
- Tuesday: Create a superhero based on your job
- Wednesday: Invent a new holiday and its traditions
- Thursday: Write a review of your day like it's a movie
- Friday: Design a fictional product that solves your pet peeve

The Creative Expansion Method Start small: "Write a joke" Build: "Make it about coffee" Expand: "For my coworkers who know I drink too much" Perfect: "Include reference to my famous 3pm crash"

The Collaboration Approach You provide: Personal details, memories, inside jokes ChatGPT provides: Structure, polish, creative angles Together: Something uniquely yours

Real People, Real Creativity

The Reluctant Artist "I always said I wasn't creative. ChatGPT helped me write songs for my guitar (three chords only!), design a garden that tells a story, and create bedtime tales my kids beg for. I'm not creative – I just needed help expressing it." - Tom, engineer

The Memory Maker "Every grandkid gets a personalized story for their birthday where they're the hero. ChatGPT helps me weave in their current interests and challenges. My daughter says these stories are their favorite gifts." - Grandma Betty

The Connection Creator "I use ChatGPT to write little poems for my wife's lunch notes. Different one each day. She's kept every single one. Our teenage daughter asked how we're still romantic after 20 years. It's the little things, with a little help." - Devoted husband

Your Creative Challenge

This week, create something for someone:

1. **Monday**: Write a thank you note that's actually memorable

2. **Tuesday**: Create a personalized joke or riddle

3. **Wednesday**: Design a simple game for family/friends

4. **Thursday**: Write a short story featuring someone you love

5. **Friday**: Craft something that makes someone laugh

The Robert Revelation

Six months after that birthday poem, Robert sent me a note: "Emma framed the poem. But more importantly, I kept writing. Little poems for my wife, funny songs about spreadsheets for office parties, stories for my nephew. I'm still an accountant. But now I'm an accountant who creates joy with words. ChatGPT didn't make me creative – it showed me I always was."

The Creative Truth

Creativity isn't about being Shakespeare or Picasso. It's about expressing what's inside you in ways that connect with others. It's about making ordinary moments memorable. It's about sharing joy, laughter, love, and meaning.

ChatGPT is your creative catalyst. It doesn't replace your creativity – it unleashes it. It turns "I wish I could" into "Look what I made." It transforms creative impulses into creative expressions.

You have stories to tell, jokes to share, beauty to create. ChatGPT is ready to help you express them all.

What will you create today?

Lesson 4.5: Real-Life Case Study – Plan a Vacation with ChatGPT

"I'm terrible at planning vacations," admitted Jennifer, scrolling through her phone with a mixture of excitement and dread. "Last year, we spent more time arguing about restaurants than actually enjoying San Francisco. My husband wants an itinerary, my teenage daughter wants 'spontaneous adventure,' and my 10-year-old just wants to swim. I want everyone to be happy without losing my mind."

"Let's plan your perfect family vacation in the next 30 minutes," I suggested, opening ChatGPT.

Jennifer laughed. "It took me three weeks to not plan our last vacation properly."

What followed was a masterclass in using ChatGPT's conversational abilities to solve complex, multi-variable problems. By the end, Jennifer had a complete vacation plan that balanced structure with flexibility, adventure with relaxation, and individual desires with family togetherness.

The Progressive Planning Approach

Instead of dumping everything into one massive prompt, we built the vacation plan layer by layer, just like having a conversation with a talented travel agent who happens to know everything.

Step 1: Setting the Foundation

Jennifer's first prompt: "Help me plan a family vacation. Budget: $5000 total. Family of 4: two adults, 15-year-old daughter, 10-year-old son. Duration: 7 days. We live in Ohio. Needs: mix of relaxation and activities, some beach time, good food. Timeframe: this July."

ChatGPT responded with clarifying questions:

- Preference for domestic or international?
- Drive or fly?
- One location or multiple stops?
- Hotel, resort, or rental property?
- Any specific interests or must-avoids?

Jennifer's response: "Domestic, willing to fly, prefer staying in one place with day trips, leaning toward rental property for space, daughter loves photography, son obsessed with marine life, husband likes history, I just want one day to read on a beach."

Step 2: Destination Deep Dive

ChatGPT suggested three destinations with rationales:

1. **Outer Banks, North Carolina**: Beach, Wright Brothers history, wild horses for photography, excellent aquarium

2. **San Diego, California**: Beaches, world-class zoo, USS Midway, perfect weather

3. **Florida Keys**: Snorkeling, beaches, Hemingway history, laid-back vibe

Jennifer: "San Diego sounds perfect! But is it doable on our budget?"

Next prompt: "Break down a realistic budget for 7 days in San Diego for a family of 4. Include: flights from Columbus, accommodation, food, activities, and transportation. Show where we can save and where to splurge."

ChatGPT provided:

- Flights: $1,600 (book Tuesday/Wednesday for savings)
- Accommodation: $1,400 (VRBO in Pacific Beach)
- Car rental: $400
- Food: $1,000 ($140/day mixing restaurants and groceries)
- Activities: $500
- Buffer: $100

Total: $5,000

Step 3: The Detailed Itinerary

"Create a 7-day San Diego itinerary that includes: beach time daily, one full relax day for me, photography opportunities for daughter, marine life for son, some history for husband. Mix must-sees with downtime. Include restaurant suggestions but also grocery shopping for breakfast/lunch at rental."

ChatGPT created:

Day 1 (Arrival):

- Land 2pm, get rental car
- Grocery shop ($200 budget, list provided)
- Check into rental in Pacific Beach
- Sunset walk on beach (photography op)
- Dinner at casual beachfront restaurant

Day 2 (Marine Life Day):

- Morning: La Jolla Cove (seals, snorkeling)
- Afternoon: Birch Aquarium
- Evening: Beach time at rental
- Dinner: Cook at rental (ChatGPT suggested easy meal)

Day 3 (History & Culture):

- Morning: USS Midway Museum
- Lunch: Little Italy
- Afternoon: Balboa Park (multiple museums)
- Evening: Sunset Cliffs (photography)
- Dinner: Mexican in Old Town

Day 4 (Mom's Beach Day):

- Family breakfast at rental
- Drop mom at beach with book
- Others: Torrey Pines hike (photography)
- Afternoon: Everyone beach time
- Dinner: BBQ at rental

Day 5 (Zoo Day):

- Full day San Diego Zoo
- Pack lunch to save money
- Dinner: Treat meal at nice restaurant

Day 6 (Adventure Day):

- Morning: Kayaking in La Jolla
- Afternoon: Coronado Beach
- Evening: Hotel del Coronado tour
- Dinner: Coronado

Day 7 (Departure):

- Morning: Last beach time
- Clean rental
- Flight home 3pm

Step 4: The Reality Check

Jennifer: "This looks amazing but maybe too packed. My family moves slowly in the mornings."

Revision prompt: "Adjust the itinerary for a family that doesn't get moving until 10am. Build in more flexibility and downtime. Mark what's essential vs. optional."

ChatGPT revised with:

- Later start times
- "Must do" vs. "If we feel like it" activities
- Backup indoor options for each day
- More beach time buffers
- Restaurant alternatives near each activity

Step 5: The Deep Dives

"Give me detailed information for Day 2 (Marine Life Day) including: exact addresses, parking tips, best time to see seals, snorkel rental locations, ticket prices, and kid-friendly restaurant for dinner."

ChatGPT provided everything down to which parking lot fills up first and where to stand for best seal photos.

Jennifer then asked for similar deep dives on:

- Restaurant recommendations with menus that work for picky eaters
- Beach comparison chart (which beach for which activity)
- Photo spots mapped for golden hour
- Rainy day backup plan

Step 6: The Preparation Lists

"Create pre-vacation preparation lists including: packing list for San Diego weather, what to book in advance, rental car tips, and travel day checklist."

ChatGPT organized everything into phone-screenshot-ready lists:

Book 2 Months Before:

- Flights (Tuesday/Wednesday)
- Rental property
- Rental car

Book 2 Weeks Before:

- Zoo tickets (online discount)

- Midway Museum (timed entry)
- One nice restaurant

Pack Smart:

- Layers (morning fog, afternoon sun)
- Reef-safe sunscreen
- Water shoes for tide pools
- Portable phone charger
- Beach toys (or buy there)

The Personalization Magic

Throughout the planning, Jennifer added personal touches: "My daughter is vegetarian" → Restaurant suggestions updated "My son gets cranky when hungry" → Snack stops added "I hate tourist traps" → Authentic local spots prioritized "We're early risers on vacation" → Sunrise beach walks added

The Results

Three Months Later - Jennifer's Report: "Best. Vacation. Ever. The plan gave us structure without strangling spontaneity. My husband loved having reservations handled. My daughter got incredible photos at every spot ChatGPT suggested. My son talks about the tide pools constantly. And I got my beach reading day."

What Made It Work:

- Built progressively, not all at once
- Balanced everyone's needs
- Mix of planned and flexible

- Practical details included
- Personalized to family dynamics

The Unexpected Benefits: "ChatGPT suggested we create a shared photo album where everyone uploads daily favorites. It became our favorite vacation tradition. Also, the grocery list was perfect – we actually used everything."

Your Vacation Planning Playbook

The Foundation Prompt: "Help me plan a [duration] vacation for [who] to [where/type]. Budget: [amount]. Interests: [list]. Constraints: [list]. Must-haves: [list]."

The Progressive Layers:

1. Destination selection with rationale

2. Budget breakdown

3. Day-by-day itinerary

4. Detailed activity information

5. Logistics and preparation

6. Personalization adjustments

The Smart Follow-Ups:

- "What am I not thinking about?"
- "Where do tourists waste money?"
- "What would locals recommend?"
- "How to handle a [common problem]?"
- "Make it work for [specific family dynamics]"

Beyond Basic Vacation Planning

ChatGPT can handle complex scenarios:

Multi-Generational Trip: "Plan for ages 5 to 75 with mobility considerations" **Adventure Travel**: "Balance adventure with safety for family" **Cultural Immersion**: "Help us experience, not just observe" **Budget Crisis**: "Flights doubled – reorganize within budget"

The Planning Principles

1. **Start broad, then narrow** - Destination → Itinerary → Details

2. **Include everyone's must-haves** - No one feels ignored

3. **Build in flexibility** - Plans are guides, not prisons

4. **Get specific when needed** - Exact addresses, prices, times

5. **Prepare for problems** - Rain plans, cranky kid strategies

Your Turn: Plan Your Perfect Trip

This week, plan a trip (real or dream):

1. **Define parameters** - Who, when, where, budget

2. **Build progressively** - Don't overwhelm with one prompt

3. **Get detailed** - Specific days, specific needs

4. **Personalize ruthlessly** - Your family, your style

5. **Prepare thoroughly** - Lists, backups, confirmations

The Vacation Victory

Jennifer's family still talks about their San Diego trip. Not because everything went perfectly – it didn't. But because the planning gave them freedom to enjoy instead of stress about decisions.

"ChatGPT didn't plan our vacation," Jennifer reflects. "We planned our vacation. ChatGPT just made us really, really good at it. It asked questions I didn't know to ask and thought of details I always forget. It turned vacation planning from a chore into part of the adventure."

The secret? ChatGPT doesn't replace human judgment about what makes a great vacation. It amplifies your ability to create one. It's the difference between hoping for a good trip and designing one.

Your perfect vacation is waiting to be planned. Where will ChatGPT help you go?

CHAPTER 5

Workplace Productivity

~~~

Opening Story: The Transformation of Marcus Chen

Marcus Chen was burning out. As a senior project manager at a tech company, he spent 60-hour weeks drowning in status reports, meeting notes, emails, and documentation. His actual strategic work? Maybe 10% of his time.

"I'm basically a highly paid administrator," he told his wife one evening, laptop still open at 9 PM. "I got into project management to solve problems and lead teams, not to format documents."

Then his colleague Lisa mentioned she'd been using ChatGPT for work. Marcus was skeptical. "Isn't that... cheating?"

Lisa laughed. "Is using Excel instead of a calculator cheating? It's a tool, Marcus. Let me show you."

One month later, Marcus's reality had shifted:

- Document creation time: Down 70%
- Email response time: Cut in half
- Meeting prep: 15 minutes instead of an hour
- Weekly reports: Automated to 10-minute reviews
- Strategic work time: Up to 40% of his week

"I'm doing the best work of my career," Marcus told me. "Not because ChatGPT does my job, but because it handles the repetitive parts so I can do the thinking parts. My boss asked what changed. I told her: "I learned to delegate to AI."

Lesson 5.1: Productivity Boosters – Work Smarter, Not Harder

The average knowledge worker spends 41% of their time on routine tasks that don't require human creativity or judgment. That's two full days every week spent on work that ChatGPT could handle in minutes. Let's reclaim that time.

The Task Audit Revolution

Before automating anything, understand where your time actually goes. Here's how Sarah, a marketing director, discovered her productivity gold mine:

Sarah's Time Audit Results:

- Writing first drafts: 8 hours/week
- Formatting and editing: 6 hours/week
- Meeting summaries: 4 hours/week
- Email responses: 10 hours/week
- Data analysis reports: 5 hours/week
- Actual strategic thinking: 7 hours/week

"I was shocked," Sarah admitted. "I'm paid for strategy, but I spend 80% of my time on tasks an assistant could do. Except I don't have an assistant. Now I do – ChatGPT."

The Email Transformation

Email is the biggest time thief in modern work. Let's turn it from burden to breeze.

The Batch Processing Method Instead of reactive responses, Marcus processes email in focused blocks:

Morning Email Sprint (30 minutes):

1. Quick scan for urgent issues

2. Copy similar emails to ChatGPT

3. Prompt: "Write professional responses to these 5 emails. Context: I'm a project manager. Tone: Friendly but efficient. Keep each under 100 words."

4. Personalize and send

Email Templates That Scale Create your response library:

"Generate 10 variations of [common email type] that I can quickly customize. Include:

- Scheduling meetings
- Requesting updates
- Declining meetings
- Providing project status
- Following up on requests Each should be professional but warm, 75-100 words."

Real Success: "I used to spend 2 hours on Monday morning just catching up on weekend emails. Now it's 30 minutes, and my responses are actually better." - Marcus

Meeting Management Mastery

The Pre-Meeting Prep Protocol Old way: Spend 45 minutes creating an agenda New way: 5-minute ChatGPT collaboration

Prompt: "Create a focused agenda for tomorrow's project status meeting. Attendees: 2 developers, 1 designer, product owner. Key issues: delayed API integration, design revision requests, timeline concerns. Duration: 30 minutes. Include time allocations and desired outcomes for each item."

The Magic of Meeting Summaries During meeting: Take rough notes After meeting: Transform into gold

Prompt: "Transform these rough meeting notes into a professional summary. Format:

- Key decisions (bullet points)
- Action items with owners and deadlines (table format)
- Open issues for next meeting
- One-paragraph executive summary at top Keep total under 300 words."

Results: "My meeting summaries went from everyone asking for clarification to everyone saying they're the clearest they've ever seen." - Team Lead Jennifer

Document Creation at Light Speed

The First Draft Freedom Never stare at blank pages again.

Project Proposals "Create a project proposal outline for [specific project]. Include:

- Executive summary

- Problem statement
- Proposed solution
- Timeline with milestones
- Resource requirements
- Success metrics
- Risk analysis Target length: 5 pages. Audience: C-suite executives."

Then fill in with your specific details.

Status Reports That Shine Weekly drudgery becomes 10-minute task:

"Based on these bullet points of this week's progress, create a professional status report: [Paste your rough notes]

Format:

- Executive summary (2 sentences)
- Achievements this week
- Challenges and solutions
- Next week's priorities
- Metrics dashboard Tone: Confident but honest about challenges"

The Analysis Accelerator

Data Storytelling Transform numbers into narratives:

"I have this sales data: [paste data]. Create:

1. Three key insights

2. What the trend suggests

3. Recommended actions

4. One visualization suggestion: Write for non-technical executives who care about growth."

Competitive Intelligence "Analyze these three competitor product descriptions. Create comparison table highlighting:

- Our advantages

- Their advantages

- Gaps we could exploit

- Positioning recommendations"

Real-World Productivity Wins

The Consultant's Transformation "I bill by the hour but felt guilty charging for routine tasks. Now ChatGPT handles document formatting, first drafts, and research summaries. I focus on strategy and client relationships. Revenue up 40%, stress down 50%." - Independent consultant

The HR Manager's Revolution "I used to spend entire days writing job descriptions and policy documents. Now I prompt ChatGPT with our requirements and company voice, then customize. What took 8 hours now takes 1. I actually have time for the 'human' part of Human Resources." - HR Director

The Sales Rep's Secret "ChatGPT helps me personalize 50 outreach emails in the time it used to take for 5. Each references the prospect's company, challenges, and includes relevant case study. Response rates tripled." - Enterprise sales rep

The Productivity System Stack

Daily Productivity Rituals

Morning Power Hour (7:30-8:30 AM):

- Email batch processing (20 min)
- Daily priority list creation (10 min)
- Document drafting for day (30 min)

Afternoon Administration (3:00-3:30 PM):

- Meeting notes cleanup
- Status updates
- Tomorrow's agenda prep

Weekly Productivity Sessions

Monday Planning (30 minutes):

- Week's priorities with ChatGPT
- Calendar optimization
- Batch create recurring documents

Friday Wrap-Up (45 minutes):

- Weekly report generation
- Email templates for next week
- Process improvements brainstorm

Advanced Productivity Techniques

The Context Switch Minimizer "I have 5 different types of tasks today. Organize them to minimize context switching. Consider:

energy levels throughout the day, task dependencies, meeting schedule [paste calendar]. Create time blocks."

The Decision Accelerator "I need to decide between Option A and Option B for [situation]. Create:

- Pros/cons for each
- Risk assessment
- Questions I should consider
- Recommendation with reasoning"

The Learning Optimizer "I need to get up to speed on [new topic] for work. Create:

- 20% of concepts that give 80% understanding
- Key terminology with simple definitions
- Three 10-minute learning sessions
- How this applies to my role as [job title]"

Productivity Pitfalls to Avoid

Don't Abdicate Thinking ChatGPT handles routine tasks, not strategic decisions. You still own the strategy.

Don't Skip Personalization Always add your voice, specific context, and human judgment.

Don't Create Without Reviewing Speed is good. Accuracy is essential. Always review before sending.

Don't Forget Security Never put confidential data into ChatGPT. Use placeholders and add sensitive details later.

Your Productivity Transformation Plan

Week 1: Email Liberation

- Create email template library
- Batch process responses
- Track time saved

Week 2: Meeting Mastery

- Automate agenda creation
- Streamline note summaries
- Optimize meeting prep

Week 3: Document Domination

- Build proposal templates
- Automate reports
- Speed up first drafts

Week 4: Analysis Acceleration

- Create data story templates
- Build comparison frameworks
- Develop insight generators

The Compound Effect

Marcus discovered something profound: "The time saved compounds. I use those reclaimed hours for strategic work, which makes my projects more successful, which gives me more influence, which lets me delegate more, which frees up more time for strategy. It's an upward spiral."

His metrics after 6 months:

- Promoted to Senior Principal PM
- Leading company's most strategic initiative
- Working 45 hours instead of 60
- Job satisfaction highest ever

"ChatGPT didn't just make me more efficient," Marcus reflects. "It let me become the professional I always wanted to be – focused on solving big problems, not drowning in small tasks."

Your Productivity Challenge

This week, pick your biggest time drain:

1. **Measure it** - How many hours per week?
2. **Prompt it** - Create ChatGPT solution
3. **Test it** - Try for three days
4. **Refine it** - Adjust based on results
5. **Scale it** - Apply to similar tasks

Remember: The goal isn't to work faster. It's to work on what matters. ChatGPT handles the routine so you can be extraordinary.

What will you do with your reclaimed time?

Lesson 5.2: Content Creation for Work – First Drafts and Beyond

Rebecca stared at her laptop screen, cursor blinking in an empty document. As Director of Communications for a healthcare startup, she needed to create:

- A 20-page white paper on patient data security
- Website copy for a new product launch
- Internal newsletter about company culture
- LinkedIn thought leadership posts
- Investor update presentation

All due by month's end. All requiring her unique expertise and company knowledge. All starting from blank pages.

"I'm a good writer," she told me, "but I spend 80% of my time staring at empty documents, trying to find the right starting point. By the time I actually write, I'm mentally exhausted."

Two hours later, after I showed her how to use ChatGPT as a content creation partner, Rebecca had solid first drafts for all five projects. "It's like having a brilliant junior writer who never sleeps and always has ideas," she marveled. "I still do the strategic thinking and final polish, but the blank page paralysis is gone."

The First Draft Revolution

The hardest part of content creation isn't perfecting – it's starting. ChatGPT eliminates blank page syndrome by giving you something to react to, refine, and build upon.

The Progressive Draft Method Instead of trying to create perfect content in one shot, build it in layers:

Layer 1: Structure "Create a detailed outline for a white paper on patient data security for healthcare IT decision makers. Include:

- Current threat landscape

- Regulatory requirements
- Best practices
- Implementation roadmap
- ROI considerations Target length: 20 pages. Tone: Authoritative but accessible."

Layer 2: Section Development "Now write the introduction section based on this outline. Hook readers with a recent healthcare data breach statistic. Establish why this matters now. Preview the value they'll get. 300 words."

Layer 3: Enhancement "Good foundation. Now add:

- A specific example from a mid-size hospital
- One compelling quote about data breach costs
- Clearer transition to next section"

Layer 4: Personalization Add your company's unique perspective, proprietary data, and specific solutions.

Content Types That Transform

Website Copy That Converts Old process: Agonize over every word for days New process: Strategic prompting in hours

"Write homepage hero section copy for our patient monitoring software. Target audience: Hospital CTOs. Key benefits: 50% reduction in false alarms, AI-powered predictions, integrates with existing systems. Tone: Confident but not salesy. Include a compelling headline and 50-word description."

Then iterate:

- "Make it more benefit-focused"
- "Add urgency without being pushy"
- "Include social proof element"

Real Result: "Our new site copy converted 3x better than what our expensive agency wrote. Because I could test 10 versions in the time they took for one." - SaaS Marketing Director

Internal Communications That Engage

The Newsletter Nobody Reads → The Newsletter Everyone Loves

Prompt progression:

1. "Create an internal newsletter outline celebrating our engineering team's product launch. Include: behind-the-scenes challenges, individual spotlights, lessons learned, what's next. Tone: Proud colleague, not corporate PR."

2. "Write the opening section with a hook about the 3am breakthrough that saved the launch"

3. "Add 5 engineering inside jokes only our team would understand"

Result: 87% open rate, up from 31%

Thought Leadership That Establishes Authority

LinkedIn Posts That Get Noticed "I'm a healthcare IT executive. Write a LinkedIn post about why hospital cybersecurity is like infection control. Draw parallels between digital and biological viruses. Include one counterintuitive insight. 200 words. End with a thought-provoking question."

Then personalize with your specific experience.

Blog Posts That Rank "Create a blog post outline: '10 Patient Data Security Mistakes Hospitals Make.' For each mistake:

- What it is
- Real-world example (anonymized)
- The fix
- Implementation tip SEO optimize for 'hospital data security.' 2000 words total."

Presentation Creation at Scale

The Executive Update Accelerator From blank slides to polished deck in 45 minutes:

"Create investor update presentation outline covering:

- Quarterly highlights (3 key metrics)
- Product development progress
- Customer wins
- Challenges and solutions
- Q4 priorities
- Ask (what we need from the board) 10 slides max. Executive-friendly."

For each slide: "Write slide content for 'Customer Wins.' Include:

- Headline stat
- 3 bullet points with specific examples

- One customer quote
- Visual suggestion"

Advanced Content Techniques

The Voice Matching Method "Here are 3 examples of our company blog posts: [paste excerpts]. Analyze the voice, tone, and style. Now write a new post about [topic] matching this exact voice."

The Repurposing Engine "Take this 20-page white paper and create:

1. Executive summary (1 page)

2. Blog post series (3 posts, 500 words each)

3. LinkedIn article (800 words)

4. Email campaign (5 emails)

5. Webinar outline (45 minutes)

6. Infographic text content"

The Localization Accelerator "Adapt this product description for:

- Technical buyers (emphasize specs)
- Business buyers (emphasize ROI)
- End users (emphasize ease of use) Keep core message consistent."

Real-World Content Victories

The Product Manager's Portfolio "I used to struggle writing feature announcements. Now I prompt ChatGPT with technical details and target benefits. It creates clear, exciting copy that engineering approves and customers understand. Ship announcements went from 2-day ordeals to 2-hour tasks." - Senior PM at Microsoft

The Startup Founder's Scale "We couldn't afford a content team. ChatGPT helped me create:

- 50 blog posts (I edited/personalized each)
- Complete website copy
- Email nurture sequences
- Sales collateral library Content that would've cost $50K for $20/month." - B2B startup founder

The Consultant's Authority Builder "I write one thoughtful LinkedIn post daily using ChatGPT for first drafts. Added 10K relevant followers in 6 months. More importantly, inbound leads up 400%. People say my content is exactly what they need to hear." - Management consultant

The Quality Control System

The Three-Pass Method

1. **ChatGPT Draft**: Get ideas flowing
2. **Human Enhancement**: Add expertise, examples, personality
3. **ChatGPT Polish**: Grammar, flow, consistency

The Fact-Check Protocol

- Never trust ChatGPT's statistics
- Verify any claims or quotes
- Add your real data and examples

- Have subject expert review

The Voice Consistency Check "Review this content. Does it match our brand voice guide:

- Professional but approachable
- Data-driven but human
- Confident without arrogance Point out any sections that need adjustment."

Content Creation Workflows

The Weekly Blog Machine Monday (30 min): Brainstorm topics with ChatGPT Tuesday (45 min): Create outlines for month Wednesday (2 hours): Write 4 first drafts Thursday (2 hours): Edit, personalize, add data Friday (1 hour): Schedule and promote

Output: Month of content in 6 hours

The Campaign Creator Start with core message → Generate all formats:

- Long-form pillar content
- Social media series
- Email sequences
- Sales enablement tools
- Paid ad copy

All consistent, all connected, all in one day.

Avoiding Content Pitfalls

Don't Publish Raw Output ChatGPT creates foundations, not finished products. Always add your expertise.

Don't Ignore SEO "Include these keywords naturally: [list]. Optimize for search intent: [description]."

Don't Forget Compliance In regulated industries, have legal/compliance review everything.

Don't Lose Authenticity Your stories, your data, your insights make content valuable. ChatGPT provides structure.

Your Content Creation System

Build Your Prompt Library:

- Blog post structures
- Email templates
- Social media formats
- Report frameworks
- Presentation outlines

Create Your Voice Guide: "Our content voice is [description]. Here are 5 examples: [paste]. Always match this style."

Develop Your Workflow:

1. Brainstorm with ChatGPT
2. Create detailed outlines
3. Generate section drafts
4. Add your expertise

5. Polish with ChatGPT

6. Human final review

The Rebecca Resolution

Six months later, Rebecca leads content strategy for her company's most successful product launch ever. "ChatGPT didn't replace my creativity – it amplified it. I spend time on strategy and storytelling, not staring at blank pages. My output quadrupled, but more importantly, quality improved because I have energy for the human elements that matter."

Her advice: "Think of ChatGPT as your infinitely patient writing partner who always has ideas and never judges your rough drafts. You bring the strategy, expertise, and soul. Together, you create content that neither could alone."

Your Content Challenge

This week, transform your content creation:

1. **Audit your needs** - List all content you create regularly

2. **Build templates** - Create prompts for each type

3. **Test the system** - Create one piece using the progressive method

4. **Refine and scale** - Adjust process, apply broadly

5. **Track results** - Measure time saved and quality impact

The blank page is dead. Long live the first draft. What will you create today?

Lesson 5.3: Skill Development – Learning on the Job with AI

David felt like a fraud. After 15 years as a successful sales manager, he'd just been promoted to VP of Sales at a tech company. The problem? Everyone kept talking about "data-driven decisions," "SQL queries," and "predictive analytics." David knew sales, not data science.

"I can't go back to school," he told me. "I have three kids and a massive new role. But if I don't figure out data analytics fast, I'll fail."

"What if you could have a personal tutor available 24/7, who teaches exactly what you need, at your pace, in context of your actual work?" I asked, opening ChatGPT.

Three months later, David was leading data strategy meetings, building dashboards, and making decisions backed by analysis he actually understood. "ChatGPT didn't just teach me data skills," he said. "It taught me how to learn anything I need, whenever I need it."

The Just-in-Time Learning Revolution

Traditional learning says: Learn everything, then apply Modern reality demands: Learn what you need, when you need it ChatGPT enables: Instant, contextual, practical learning

The key shift: Stop trying to learn entire subjects. Start learning specific skills for specific tasks.

The Learn-by-Doing Method

David's data analytics journey shows the power of contextual learning:

Week 1: Foundation Building "I'm a sales leader who needs to understand data analytics. Create a learning plan that:

- Focuses only on what I need for sales analytics
- Uses sales examples throughout
- Can be done in 30 minutes daily
- Starts with absolute basics
- Builds to practical application"

ChatGPT created a custom curriculum:

- Day 1-3: What data tells us about sales
- Day 4-7: Reading basic charts and metrics
- Week 2: Understanding sales dashboards
- Week 3: Asking the right data questions
- Week 4: Basic data manipulation

The Contextual Learning Advantage Instead of generic tutorials, every example used sales data: "Explain what a conversion rate is using this example: We had 1000 website visitors, 100 filled our contact form, 25 became qualified leads, and 5 bought. Walk me through calculating and interpreting each conversion rate."

Skill Development Strategies by Domain

Technical Skills for Non-Technical People

Excel Mastery in Context Marketing manager Emma needed advanced Excel for campaign analysis:

"I need to learn Excel pivot tables for marketing analytics. Current skill: basic formulas. Create a hands-on tutorial using this sample data: [paste marketing data]. Show me:

1. What pivot tables do in plain English

2. Step-by-step creation process

3. 5 specific marketing insights I can pull

4. Common mistakes to avoid"

Result: "I went from Excel-phobic to creating dashboards my CEO loves. Learning with my actual data made it click." - Emma

Code Understanding (Not Writing) Product manager Alex worked with developers but couldn't read their updates:

"Help me understand what developers mean when they discuss our product. I don't need to code, just comprehend. Start with this actual pull request description: [paste]. Explain:

- What changed in human terms
- Why it matters to users
- What questions I should ask
- Red flags to watch for"

Business Skills with Immediate Application

Financial Literacy for Non-Finance Roles "I'm a creative director who needs to understand budgets and P&L statements. Using this simplified version of our department's financials: [paste data]

- Explain each line item in plain English
- Show what I should monitor monthly

- Highlight what affects my decisions

- Create 5 questions I should ask in budget meetings"

Strategic Thinking Development "I'm moving from individual contributor to team lead. Help me develop strategic thinking using my actual project as an example. Current situation: [describe]. Walk me through:

- Thinking beyond tasks to outcomes

- Identifying strategic opportunities

- Communicating strategically

- Common strategic thinking frameworks"

The Micro-Learning System

The 15-Minute Morning Upgrade Before checking email, invest in growth:

Monday: Learn one new Excel function for your work Tuesday: Understand one business concept you encountered Wednesday: Improve one communication skill Thursday: Master one productivity technique Friday: Explore one industry trend

Example Monday session: "Teach me Excel's VLOOKUP function using sales data. Show me:

- What problem it solves

- Step-by-step with my data

- 3 ways I'd use it this week

- Common errors to avoid: Keep it under 15 minutes."

Real-World Skill Development Wins

The Designer Who Learned Business "As a designer, I was always excluded from strategy discussions. ChatGPT helped me understand:

- Business model basics using our company as example
- Financial metrics that matter to executives
- How design impacts business outcomes
- Strategic frameworks in design context

Now I'm Lead Designer because I connect design to business value." - Senior Designer at Spotify

The Engineer Who Mastered Communication "My code was solid but my presentations were painful. ChatGPT became my communication coach:

- Simplified my technical explanations
- Taught me storytelling structures
- Helped prepare for specific meetings
- Practiced handling tough questions

Got promoted to Principal Engineer within 6 months." - Software engineer

The Manager Who Conquered Data "Everyone talked about being 'data-driven' but nobody taught me how. With ChatGPT:

- Learned SQL basics using our actual database
- Understood statistics through our metrics
- Built dashboards that matter

- Made first data-backed strategic decision

My team's performance improved 40% because I finally understood what to measure." - Operations manager

Advanced Learning Techniques

The Teach-Back Method "I just learned about [concept]. Let me explain it back to you. Correct any misunderstandings and suggest what I should learn next."

The Problem-First Approach "I need to solve this specific problem: [describe]. Teach me only the skills needed to solve it. We'll expand from there."

The Analogical Learning "I understand [familiar concept] well. Explain [new concept] by comparing it to what I already know. Point out where the analogy breaks down."

The Mistake-Based Learning "I tried to [task] and got this error/problem: [describe]. Explain what went wrong and teach me the right approach."

Building Your Personal Learning System

Create Your Skill Gap Map "Based on this job description for my role: [paste]. And my current skills: [list]. Identify:

- Top 5 skill gaps holding me back
- Which would have biggest immediate impact
- Learning path for each
- How to demonstrate proficiency"

Design Your Learning Rhythm

- Daily: One micro-skill (15 minutes)

- Weekly: One deeper concept (1 hour)

- Monthly: One major capability (project-based)

Track Your Progress "I've been learning [skill] for two weeks. Test my understanding with 5 questions ranging from basic to advanced. Based on my answers, suggest next steps."

The Meta-Skill: Learning How to Learn

ChatGPT's greatest gift isn't teaching specific skills – it's teaching you how to learn anything:

Identify What You Actually Need "I think I need to learn [broad subject]. Given my role as [position] and these challenges: [list], what specific aspects should I focus on?"

Find Your Learning Style "Explain [concept] three ways:

1. Visual/diagram-based

2. Story/example-based

3. Logical/step-by-step I'll tell you which clicked best."

Create Practice Opportunities "I just learned [skill]. Create 5 practice scenarios increasing in difficulty, based on situations I'd actually encounter at work."

Your Skill Development Action Plan
Week 1: Assess and Plan

- List skills you need but don't have

- Prioritize by immediate impact
- Create learning plan for top skill

Week 2: Daily Practice

- 15 minutes each morning
- Apply to real work same day
- Track what works

Week 3: Deepen and Expand

- Tackle more complex aspects
- Connect to existing knowledge
- Teach someone else

Week 4: Integrate and Systematize

- Make learning routine
- Build your prompt library
- Plan next skill

The David Transformation

One year later, David isn't just surviving as VP of Sales – he's thriving. "The data skills were just the beginning. ChatGPT taught me that I can learn anything. New CRM system? Two weeks. Presentation design? One week. Board reporting? Three days. I'm not smarter – I just learned how to learn."

His advice: "Stop thinking about learning as events (courses, degrees). Start thinking about learning as a continuous conversation with the smartest tutor you'll ever have. Every gap is just a question away from being filled."

The Learning Mindset Shift

Traditional view: "I don't know that, so I can't do that job" ChatGPT-enabled view: "I don't know that yet. Give me a week"

The workplace increasingly rewards those who can learn quickly, not those who learned long ago. ChatGPT makes you infinitely upgradeable.

Your career isn't limited by what you know. It's expanded by how fast you can learn what you need to know. And with ChatGPT, that speed approaches real-time.

What skill will you master this week?

Lesson 5.4: Business Considerations – Using AI Responsibly at Work

The conference room fell silent when Lisa finished her presentation. As Chief Legal Officer, she'd just shown how a competitor was sued for $2.3 million after an employee accidentally shared confidential client data with ChatGPT, which then appeared in responses to other users.

"So we're banning it?" asked Tom from Sales, his face falling. He'd been using ChatGPT to write proposals.

"No," Lisa replied. "We're going to use it intelligently. The companies getting in trouble are the ones with no policies. The ones thriving have clear guidelines. Today, we create ours."

What followed was a masterclass in balancing innovation with responsibility, productivity with protection. Six months later, their company was a case study in responsible AI use – more productive than ever, without a single incident.

The Business Reality Check

Here's the truth: Your employees are already using AI. Studies show 68% of knowledge workers have tried ChatGPT for work tasks. The choice isn't whether AI enters your workplace – it's whether it enters responsibly or recklessly.

The Three Categories of Companies:

1. **The Deniers**: Ban AI, fall behind, lose talent to progressive competitors

2. **The Reckless**: No guidelines, eventual data breach or compliance failure

3. **The Smart**: Clear policies enabling safe innovation

Let's build your path to category three.

Creating Your AI Usage Policy

The Foundation Framework Lisa's team created a policy covering six pillars:

1. Approved Use Cases Clear green lights for:

- First drafts of non-confidential content
- Brainstorming and ideation
- Learning and skill development

- Code debugging (non-proprietary)
- General research and analysis

2. Prohibited Uses Absolute red lines:

- Client confidential information
- Employee personal data
- Proprietary code or algorithms
- Financial data or projections
- Strategic plans or trade secrets
- Legal or medical advice

3. The Data Classification System

- **Public**: Press releases, marketing content → Safe for AI
- **Internal**: Policies, general communications → Use with caution
- **Confidential**: Client data, financials → Never use AI
- **Secret**: Trade secrets, M&A plans → Absolutely prohibited

Real Implementation: "We created a simple traffic light system. Green data = AI-friendly. Yellow = scrub identifying info first. Red = never touch AI. Employees get it instantly." - Tech company CISO

Practical Safety Protocols

The Scrubbing Method Before using ChatGPT for sensitive tasks:

Original: "Help me analyze why Acme Corp's campaign failed" Scrubbed: "Help me analyze why a B2B software campaign targeting CFOs failed"

Original: "Review John Smith's performance review" Scrubbed: "Review a senior developer's performance review"

The Placeholder Protocol Sales team's approach: "We use [CLIENT] instead of names, [PRODUCT] for our solutions, [COMPETITOR] for competition. After ChatGPT helps with structure, we replace it with real details."

The Review Requirement "All AI-assisted content for external use must be reviewed by humans before sending. Period. This caught three instances where ChatGPT invented plausible-sounding but false statistics." - Marketing Director

Industry-Specific Considerations

Healthcare: HIPAA Compliance "We use ChatGPT for administrative tasks, never patient data. Example: 'Write patient discharge instructions for common procedure' not 'Write discharge instructions for John Doe's surgery.'" - Hospital administrator

Financial Services: Regulatory Requirements "Our compliance team created AI-safe templates. ChatGPT helps with structure and clarity, never specific client portfolios or recommendations." - Investment firm CCO

Legal: Privilege and Confidentiality "We use AI for research and drafting templates, never client-specific work. It's like having a very smart law clerk who only knows general law." - Law firm partner

Education: FERPA and Academic Integrity "Teachers use ChatGPT for lesson planning and general feedback templates, never with actual student work or grades." - School district policy

The ROI of Responsible AI Use

Productivity Gains Without Risk Company case study (500 employees):

- 30% reduction in document creation time
- 50% faster first drafts
- 25% improvement in writing quality
- Zero data breaches
- Zero compliance violations

"The key was being specific about what's allowed. Employees feel safe experimenting within boundaries." - CEO

Building AI Literacy Across Teams

The Training Program That Works

Week 1: Basics and Benefits

- What AI can and can't do
- Safe use demonstrations
- Hands-on practice with approved tasks

Week 2: Security and Risks

- Data classification review
- Scrubbing techniques

- Real breach examples

Week 3: Advanced Applications

- Team-specific use cases
- Productivity workflows
- Quality control processes

Week 4: Ongoing Excellence

- Updates on new capabilities
- Sharing best practices
- Continuous improvement

Managing the Human Side

Addressing Fear: "AI Will Replace Me" "We positioned AI as a career accelerator, not replacement. Employees who embrace it get promoted faster because they accomplish more strategic work." - HR Director

Handling Resistance: "This Is Too Complicated" "Start with one simple use case per team. Sales started with email templates. Once they saved 5 hours/week, everyone wanted to learn more." - Change management consultant

Encouraging Innovation: "What Else Could We Do?" "Monthly AI innovation sessions where teams share new uses they've discovered. Best ideas get implemented company-wide." - Innovation officer

Real-World Policy Successes

The Marketing Agency's Framework "Our policy: Use AI for ideation and structure, never for final client deliverables without disclosure. Result: 3x more campaign options presented, higher client satisfaction, complete transparency." - Agency owner

The Software Company's Approach "Developers use AI for debugging and documentation, never for core product code. Productivity up 40%, intellectual property protected." - CTO

The Consulting Firm's Balance "Consultants use AI for research and framework development, add client-specific insights manually. Delivering more value faster while maintaining confidentiality." - Managing partner

Your Implementation Roadmap
Month 1: Foundation

- Form AI committee (legal, IT, HR, operations)
- Draft initial policy
- Identify pilot groups
- Create training materials

Month 2: Pilot

- Test with willing teams
- Gather feedback
- Refine policies
- Document successes

Month 3: Rollout

- Company-wide training
- Clear communication
- Support systems
- Monitoring tools

Ongoing: Evolution

- Monthly policy reviews
- Quarterly training updates
- Annual policy overhaul
- Continuous improvement

The Compliance Checklist

Before implementing AI, ensure:

☐ **Legal review** of terms of service ☐ **Data classification** system in place ☐ **Training program** developed ☐ **Incident response** plan created ☐ **Monitoring system** established ☐ **Update process** defined ☐ **Success metrics** identified

Common Pitfalls and Solutions

Pitfall: Overly restrictive policies **Solution**: Start conservative, loosen based on success

Pitfall: No enforcement mechanism **Solution**: Regular audits and clear consequences

Pitfall: Ignoring shadow IT **Solution**: Provide approved tools, understand why people seek alternatives

Pitfall: Set-and-forget policies **Solution**: Monthly reviews as AI evolves rapidly

Your AI Policy Template Starter

"As [Company], we embrace AI as a productivity tool while protecting our stakeholders.

Approved Uses: [List specific examples]

Prohibited Uses: [List specific examples]

Data Handling: [Classification system]

Review Requirements: [Who reviews what]

Training Obligations: [Who needs what training]

Violation Consequences: [Clear escalation path]

This policy evolves with technology and experience. Questions? Contact [designated person]."

The Lisa Success Story

One year later, Lisa's company is an industry leader in AI adoption. "We have zero incidents and massive productivity gains. The secret? We didn't try to stop the AI wave – we learned to surf it safely."

Key metrics:

- 35% productivity improvement
- 100% compliance maintained
- 85% employee satisfaction with AI tools
- 50% reduction in routine task time

"Responsible AI use isn't about restriction – it's about enablement with intelligence. Our employees innovate freely within clear boundaries. That's the sweet spot."

Your Action Items

This week:

1. **Assess current state** - Survey who's using what

2. **Draft basic policy** - Start simple, iterate

3. **Identify champions** - Find enthusiastic early adopters

4. **Plan pilot program** - Small group, clear metrics

5. **Communicate clearly** - This enables, not restricts

The Bottom Line

AI in the workplace isn't a question of if, but how. Companies that create thoughtful policies gain competitive advantage. Those that ignore it face risk or irrelevance.

Your choice: Be the company that fearfully restricts or intelligently enables. The tools are here. The benefits are clear. The risks are manageable.

What will your AI story be?

Lesson 5.5: AI as a Coworker – Knowing When (and When Not) to Use AI

The customer service team at TechFlow was in crisis. Call times were skyrocketing, customer satisfaction was plummeting, and agents were burning out. Sarah, the new Head of Customer

Success, had an idea: "What if we use ChatGPT to help agents respond faster?"

Two weeks later, disaster. A customer tweeted a screenshot of their chat where the agent (using ChatGPT) had confidently provided completely wrong refund information. It went viral. #TechFlowFail was trending.

But here's the plot twist: Sarah didn't abandon AI. Instead, she learned to use it correctly. Six months later, TechFlow's customer service wins awards. Response time down 50%, satisfaction up 40%, and agents actually enjoy their jobs.

The difference? Understanding when AI is your best coworker and when it's a liability.

The AI Collaboration Spectrum

Think of AI not as a replacement but as a coworker with specific strengths and weaknesses:

AI Strengths:

- Never tired or moody
- Infinite patience
- Vast general knowledge
- Lightning-fast processing
- Perfect consistency

AI Weaknesses:

- No real understanding
- No emotional intelligence
- No company-specific knowledge

- Can confidently hallucinate
- No ethical judgment

The key is knowing which tasks play to which strengths.

The Task Evaluation Framework

For every task, ask four questions:

1. Does this require real-time accuracy?

- Stock prices? ✖ AI (unless connected to live data)
- Email templates? ✔ AI perfect

2. Does this need emotional intelligence?

- Firing someone? ✖ Human only
- Drafting update email? ✔ AI with human review

3. Does this involve confidential data?

- Client strategy? ✖ Human only
- General industry analysis? ✔ AI helpful

4. Are the stakes high if wrong?

- Legal advice? ✖ Human expert only
- Blog post ideas? ✔ AI great for brainstorming

Real-World Task Sorting

Customer Service: The TechFlow Transformation

Where AI Shines:

- Drafting response templates
- Suggesting troubleshooting steps
- Finding relevant documentation
- Translating customer messages
- Summarizing long conversations

Where Humans Lead:

- Angry customer de-escalation
- Complex refund decisions
- Sensitive account issues
- Building customer relationships
- Making exception decisions

Sarah's winning formula: "AI drafts initial responses based on our templates. Agents personalize, verify accuracy, and add the human touch. AI handles structure, humans handle soul."

Department-Specific Guidelines

Sales: The Partnership Model

AI Handles:

- First draft proposals
- Competitive analysis frameworks
- Email follow-up templates

- Meeting prep summaries
- CRM data analysis

Humans Handle:

- Relationship building
- Negotiation strategy
- Reading client emotions
- Closing deals
- Strategic account planning

Success story: "I use ChatGPT to prepare for every call - company research, potential pain points, question suggestions. But once I'm on the call, it's all me. AI makes me prepared, not robotic." - Enterprise sales rep

Marketing: The Creative Collaboration

AI Excels At:

- Content ideation
- SEO optimization
- A/B test variations
- Campaign structure
- Performance analysis

Humans Own:

- Brand voice authenticity
- Creative vision
- Emotional resonance

- Cultural sensitivity
- Strategic direction

"ChatGPT gives me 50 headline options in 5 minutes. I pick the one that feels right for our brand and refine it. It's like having a tireless intern who's read every marketing book ever." - CMO

HR: The Delicate Balance

AI Appropriate:

- Job description templates
- Policy draft creation
- Benefits explanations
- Onboarding checklists
- Training materials

Human Essential:

- Performance reviews
- Conflict resolution
- Termination decisions
- Culture building
- Career counseling

"I use AI to handle paperwork so I can focus on people work. It drafts the handbook, I add the heart." - HR Director

The Decision Tree for AI Use

When facing a new task, follow this flow:

1. **Is it routine and repetitive?** → AI candidate

2. **Does it require current information?** → Verify AI limitations

3. **Are there emotional/relationship elements?** → Human leads

4. **What if AI gets it wrong?** → Assess risk level

5. **Can a human review/improve?** → Green light with oversight

Common Mistakes and Fixes

Mistake 1: The Set-and-Forget Using AI output without review

Fix: Always have human verification step

Example: Marketing team publishes AI-written blog post without review. Contains outdated statistics. Now: Human fact-checks every piece.

Mistake 2: The Emotional Outsource Using AI for sensitive communications

Fix: AI for structure, human for sentiment

Example: Manager uses ChatGPT to write layoff email. Comes across as cold. Now: AI helps with structure, manager adds empathy.

Mistake 3: The Confidentiality Breach Sharing sensitive data with AI

Fix: Scrub all identifying information

Example: Legal team uploads client contract to ChatGPT. Now: They use templates with [CLIENT] placeholders.

Mistake 4: The Over-Reliance Losing skills by always using AI

Fix: Use AI to enhance, not replace, thinking

Example: Junior analysts only use ChatGPT for reports, and can't analyze without it. Now: Required to explain AI's reasoning.

Building Your AI Collaboration System
Daily Standup Questions:

- What tasks took too long yesterday?
- Which could AI have helped with?
- Which needed human touch?
- What did we learn?

Weekly Team Review:

- Share AI wins and fails
- Update best practices
- Identify new use cases
- Refine boundaries

Monthly Evolution:

- Review AI policy compliance
- Assess productivity gains
- Identify skill gaps
- Plan training needs

Success Metrics That Matter
Quantity Metrics:

- Time saved per task

- Output volume increase
- Cost per unit work

Quality Metrics:

- Error rates
- Customer satisfaction
- Employee satisfaction
- Innovation indicators

Balance Metrics:

- Human skill development
- AI appropriate use rate
- Compliance adherence

Real Teams, Real Results

The Engineering Team "We use AI for documentation, debugging assistance, and code reviews. Never for architecture decisions or security-critical code. Result: 40% faster feature delivery, maintained code quality." - Engineering Manager at startup

The Finance Team

"AI helps with report generation and trend analysis. Humans make all decisions and verify all numbers. We catch errors AI might miss and AI catches patterns we might miss. Perfect partnership." - CFO

The Creative Team "AI is our brainstorming partner and first-draft generator. But every piece of creativity that ships has a human heart and soul. We're more creative because we spend less time on logistics." - Creative Director

Your AI Integration Playbook

Week 1: Observation Track all tasks and categorize:

- Perfect for AI
- AI with human review
- Human only

Week 2: Experimentation Test AI on low-risk tasks Document what works Note what doesn't

Week 3: Integration Build AI into workflows Create review processes Train team members

Week 4: Optimization Refine based on results Expand successful uses Eliminate problematic ones

The Future of Human-AI Collaboration

"The best teams don't see AI as us-versus-them but as us-plus-them. We're not being replaced, we're being amplified. The key is being intentional about who does what best." - Sarah, whose team now leads the industry

Her formula for success:

- AI handles volume, humans handle nuance
- AI provides options, humans make decisions
- AI ensures consistency, humans add personality
- AI saves time, humans invest it wisely

Your Collaboration Challenge

This week:

1. List your top 10 recurring tasks

2. Rate each for AI suitability (1-10)

3. Test AI on the highest-rated task

4. Document the results

5. Share learnings with your team

Remember: The goal isn't maximum AI use. It's optimal AI use. Like any coworker, AI is brilliant at some things and terrible at others. The teams that thrive are those that play to everyone's strengths – human and artificial alike.

The future of work isn't human or AI. It's human and AI, each doing what they do best. Master this collaboration, and you master the future of productivity.

What task will you delegate to your AI coworker today?

CHAPTER 6

Future of AI & Next Steps

~

Opening Story: The Conversation That Changed Everything

Maria Rodriguez had just finished presenting her company's five-year strategic plan when the board member asked the question that stopped her cold: "This is impressive, but how does AI change all of this?"

Maria, a successful CEO who'd built her logistics company from scratch, realized she'd been planning for a future that was already obsolete. Her carefully crafted projections assumed a world where AI remained a neat productivity tool, not the transformative force reshaping entire industries.

"I went home that night and spent hours with ChatGPT, not asking it to write emails, but exploring what AI might mean for logistics, for business, for society," Maria told me. "That's when I understood: we weren't just adopting a new tool. We were navigating a fundamental shift in how work, creativity, and problem-solving happen."

Six months later, Maria's company wasn't just using AI – they were building AI-first processes that put them years ahead of competitors. But more importantly, Maria had developed what she calls "AI-adaptive thinking" – the ability to continuously evolve with technology rather than being disrupted by it.

"The future isn't about predicting which AI tool will win," Maria reflects. "It's about becoming the kind of person and building the kind of organization that thrives regardless of how AI evolves."

Lesson 6.1: The Evolving Future – Trend Watch

Dr. James Chen, AI researcher at Stanford, showed me two images. The first: DALL-E's first attempt at generating "a cat" in 2021 – a blurry, vaguely cat-shaped blob. The second: A photorealistic cat from 2024's AI that fooled professional photographers.

"Three years," he said. "That's how fast AI is evolving. Not decades. Years. Sometimes months."

This breakneck pace of change means that predicting specific AI futures is foolish. Instead, smart individuals and organizations focus on understanding the patterns of change and developing adaptive capabilities.

The Velocity of Change

To understand where AI is heading, we need to grasp the speed of transformation:

2020: GPT-3 amazes researchers with coherent text **2022**: ChatGPT reaches 100 million users in 2 months **2023**: AI passes bar exam, medical licensing tests **2024**: Multimodal AI understands text, images, code, speech **2025**: AI agents beginning to work autonomously on complex tasks

"We're not on a linear progression," explains Dr. Chen. "Each breakthrough enables five more. It's exponential, and humans are terrible at understanding exponential change."

The Three Waves of AI Transformation

Wave 1: Individual Augmentation (Now - 2025) What we're experiencing today:

- Personal productivity enhancement
- Creative assistance
- Learning acceleration
- Decision support

Real example: "Our sales team uses AI for everything from email drafts to strategy planning. Each person is literally 3x more productive." - Sales Director, Fortune 500 company

Wave 2: Process Revolution (2025 - 2027) What's beginning to emerge:

- End-to-end workflow automation
- AI agents handling complete tasks
- Cross-functional AI coordination
- Predictive business operations

Early adopter: "We're piloting AI that doesn't just help with customer service – it handles entire customer journeys, escalating to humans only for complex emotional situations." - Tech startup COO

Wave 3: Structural Transformation (2027 - 2030) What futurists see coming:

- New business models enabled by AI
- Jobs that don't exist today

- AI-first organizational structures
- Human-AI collaborative entities

Visionary perspective: "Imagine a company where AI handles all routine operations, and humans focus entirely on innovation, relationships, and creative problem-solving. That's not sci-fi – that's later this decade." - Venture capitalist

Emerging Capabilities to Watch

1. Multimodal Mastery Current: ChatGPT understands text Emerging: AI that seamlessly works across text, images, video, audio, code, and data

Impact example: "Our new AI doesn't just read reports – it watches our assembly line videos, listens to machine sounds, analyzes sensor data, and predicts maintenance needs. It's like having an expert who never sleeps." - Manufacturing plant manager

2. Autonomous Agents Current: AI responds to prompts Emerging: AI that takes initiative, manages projects, coordinates with other AIs

Early implementation: "I gave our AI agent a goal: reduce customer churn by 10%. It analyzed data, created hypotheses, designed experiments, implemented changes, and measured results. All I did was approve its recommendations." - SaaS founder

3. Emotional Intelligence Current: AI mimics empathy through patterns Emerging: AI that reads emotional cues and responds appropriately

Healthcare pilot: "Our AI doesn't just track patient symptoms – it notices emotional patterns, alerts us to depression risks, and adjusts its communication style based on patient mood." - Hospital administrator

4. Creative Partnership Current: AI assists with creative tasks Emerging: AI as true creative collaborator

Artist's experience: "My AI doesn't just generate images I describe. It suggests directions I hadn't considered, builds on my ideas, and helps me explore styles I couldn't imagine alone." - Digital artist

Industry-Specific Futures

Healthcare: The AI Physician's Assistant

- Diagnosis assistance with 99%+ accuracy
- Personalized treatment plans based on genetic data
- Real-time health monitoring and prediction
- Drug discovery accelerated 100x

"We're not replacing doctors. We're giving them superpowers. Imagine diagnosing rare diseases as easily as common colds." - Medical AI researcher

Education: The Infinite Personal Tutor

- Truly personalized learning paths
- Real-time comprehension assessment
- Adaptive content generation
- Skill verification through practical application

"Every student will have an AI tutor that knows exactly how they learn best, never gets frustrated, and is available 24/7." - Education technology expert

Finance: The Predictive Money Manager

- Micro-personalized financial advice
- Real-time fraud prevention
- Automated investment optimization
- Predictive cash flow management

"Imagine an AI CFO for every small business, providing Fortune 500-level financial intelligence for $50/month." - Fintech founder

Creative Industries: The Augmented Artist

- AI co-creators in music, film, writing
- Infinite content personalization
- Real-time audience feedback integration
- New art forms we can't yet imagine

"AI isn't killing creativity – it's exploding it. We can now create things that were physically impossible before." - Film director

Preparing for Unknown Unknowns

The challenge: How do you prepare for a future when AI capabilities are evolving faster than our ability to imagine them?

The Adaptive Mindset Maria's approach: "I stopped trying to predict what AI will do. Instead, I focus on becoming someone who can quickly adapt to whatever it becomes."

Key principles:

1. **Learn continuously** - Not specific tools, but how to learn new tools

2. **Stay curious** - Every AI advancement is an opportunity

3. **Experiment freely** - Low-cost failure leads to high-value insights

4. **Network actively** - Learn from others' AI experiments

5. **Think systemically** - How does this change everything else?

The Skills That Survive

What human capabilities become MORE valuable as AI advances?

1. Ethical Judgment "AI can tell you what you could do. Only humans can decide what you should do." - Ethics professor

2. Emotional Connection "The more AI handles transactions, the more humans crave real connection." - Customer experience expert

3. Creative Vision "AI can generate a million variations. Humans decide which one matters." - Creative director

4. Complex Problem Solving "AI solves defined problems brilliantly. Humans define which problems are worth solving." - Strategy consultant

5. Adaptive Leadership "Leading through AI transformation requires skills no AI can replicate: vision, inspiration, and the ability to help others navigate change." - Executive coach

Your Future-Readiness Audit

Rate yourself (1-10) on:

- Comfort with rapid technological change
- Ability to learn new tools quickly
- Openness to AI collaboration
- Understanding of AI capabilities/limitations
- Network of AI-aware professionals
- Experimental mindset
- Ethical framework for AI use

Scores below 7? Those are your development areas.

The Practical Futurist's Toolkit
Daily Habits:

- Spend 10 minutes exploring new AI capabilities
- Ask "How might AI change this?" about routine tasks
- Share one AI discovery with colleagues

Weekly Practices:

- Test one new AI tool or feature
- Read one article about AI advancement
- Discuss AI implications with your team

Monthly Rituals:

- Attend an AI-focused webinar or event
- Experiment with AI in new area of work
- Update your AI strategy based on learnings

Quarterly Reviews:

- Assess how AI has changed your industry
- Identify emerging opportunities/threats
- Adjust skills development plan
- Network with AI pioneers in your field

Real-World Future Preparation

The Law Firm's Evolution "We created an 'AI Futures Committee' that meets monthly to explore how AI might change legal practice. We've already identified three new service areas that didn't exist last year." - Managing Partner

The Retailer's Transformation "Instead of fearing AI will replace retail workers, we're training them to become 'AI-Enhanced Customer Experience Specialists.' They use AI to provide superhuman service." - Retail CEO

The Educator's Adaptation "I redesigned my curriculum assuming every student has access to AI. Now I teach critical thinking and verification skills, not memorization." - University professor

Your Future Action Plan

This Month: Foundation

- Complete the future-readiness audit
- Join one AI-focused community
- Start daily AI capability exploration

Next Quarter: Experimentation

- Test AI in three new work areas
- Share learnings with your team

- Identify one future opportunity

This Year: Transformation

- Develop AI-adaptive workflows
- Build network of AI innovators
- Create competitive advantage through AI

The Maria Rodriguez Revelation

Remember Maria from our opening? Her company now runs "Future Fridays" where teams explore emerging AI capabilities and brainstorm applications. They've launched three new AI-enabled services and attracted top talent excited about working at the cutting edge.

"The board member who asked that question did me a favor," Maria reflects. "He forced me to stop planning for yesterday's future and start creating tomorrow's present. The key insight? The future of AI isn't something that happens to you – it's something you actively shape."

The Only Prediction That Matters

Here's the only future prediction I'm confident about: AI will continue to evolve at a pace that surprises us. The winners won't be those who predict the future correctly – they'll be those who adapt to it quickly.

Your competitive advantage isn't knowing what AI will become. It's becoming someone who thrives regardless of what AI becomes.

The future is being written in real-time. What role will you play in writing it?

Lesson 6.2: Beyond ChatGPT – A Glimpse into Other AI Tools

"I thought ChatGPT was AI," admitted Kevin, a marketing manager who'd been proudly using ChatGPT for six months. Then his colleague showed him how she was using Claude for deep analysis, Midjourney for visual content, and Perplexity for research. "It was like discovering I'd been using just one app on my smartphone while ignoring everything else."

Kevin's revelation reflects a common misconception: equating ChatGPT with all of AI. While ChatGPT opened the door, it's just one room in a vast mansion of AI capabilities. Let's explore the neighborhood.

The AI Ecosystem Map

Think of AI tools as specialists in a hospital:

- ChatGPT: The general practitioner
- Claude: The thoughtful specialist
- Google Gemini: The connected researcher
- Perplexity: The fact-checker
- Midjourney: The visual artist
- GitHub Copilot: The programming partner

Each excels in specific areas. Master contractors don't use one tool for every job – neither should you with AI.

Claude: The Thoughtful Alternative
What Makes Claude Different:

- Larger context window (can handle longer documents)

- More cautious about facts (admits uncertainty)
- Excels at analysis and reasoning
- Different "personality" - more academic

Real-World Comparison: Marketing director Sarah tested both: "I use ChatGPT for creative brainstorming and quick drafts. But for analyzing our 50-page market research report, Claude was dramatically better. It caught nuances ChatGPT missed."

When to Use Claude:

- Long document analysis
- Complex reasoning tasks
- When accuracy matters more than creativity
- Academic or technical writing
- Ethical or sensitive topics

Practical Example: "Analyze this 30-page contract and identify potential risks for a small business. Focus on payment terms, liability, and termination clauses. Explain in plain English."

Claude's response included subtleties about implied obligations that ChatGPT overlooked.

Google Gemini: The Connected Intelligence

Gemini's Superpower: Integration with Google's ecosystem

- Real-time web access
- Integration with Google Workspace
- Visual understanding capabilities

- Seamless with Android devices

The Game Changer: "Gemini can see my Google Calendar, Gmail, and Drive. When I ask 'What should I focus on today?', it actually knows my meetings, deadlines, and recent email urgencies." - Project manager

When Gemini Wins:

- Need current information
- Working within Google ecosystem
- Collaborative team projects
- Mobile-first workflows
- Visual analysis tasks

Power User Tip: "Analyze the attached spreadsheet and create a presentation about our Q3 performance. Use our brand colors from the logo in Drive, and schedule it for Monday's meeting on my calendar."

Perplexity: The Research Revolution
What Sets Perplexity Apart:

- Provides sources for every claim
- Real-time web searching
- Academic-quality citations
- Fact-checking built in

Researcher's Dream: "ChatGPT gives me answers. Perplexity gives me answers with receipts. For my PhD dissertation, that's non-negotiable." - Graduate student

Perfect For:

- Academic research
- Fact-checking important claims
- Current events analysis
- Competitive intelligence
- Source-heavy writing

Usage Example: "What are the latest developments in quantum computing for financial modeling? Include only peer-reviewed sources from 2023-2024."

Perplexity provides numbered sources you can verify.

Midjourney: The Visual Virtuoso

Beyond Text: AI image generation

- Marketing visuals
- Concept art
- Product mockups
- Presentation graphics
- Social media content

Creative Director's Take: "We used to spend $5,000 on stock photos monthly. Now we create exactly what we envision for

$30/month. More importantly, our visuals are unique." - Agency owner

Practical Applications:

- "Design a modern office space with plants and natural light"
- "Create a logo for an eco-friendly coffee shop"
- "Visualize data showing growth trends in abstract art style"

Integration Hack: Use ChatGPT to write detailed image prompts for Midjourney: "Help me describe a futuristic classroom for an AI image generator. Include specific details about lighting, colors, and atmosphere."

Specialized AI Tools by Function

For Coding: GitHub Copilot "It's like pair programming with someone who's read every programming book ever. Cuts my coding time by 40%." - Software developer

When to use:

- Writing boilerplate code
- Learning new languages
- Debugging assistance
- Documentation generation

For Music: Suno & Udio "I needed background music for my podcast. AI created custom tracks that fit perfectly. No copyright issues, perfect length." - Podcaster

Applications:

- Custom background music
- Jingles for marketing
- Mood music for videos
- Creative experimentation

For Data Analysis: Julius "Like ChatGPT specifically trained for data science. Uploads spreadsheets, creates visualizations, runs statistical analyses." - Data analyst

Best for:

- Complex data analysis
- Statistical modeling
- Automated reporting
- Predictive analytics

For Video: Synthesia & D-ID "We create training videos in 15 languages without filming anything. AI avatars deliver our content perfectly every time." - L&D manager

Use cases:

- Training content
- Multilingual communication
- Consistent messaging
- Scale video production

The Integration Strategy

Don't Replace - Complement Kevin's evolved workflow:

1. Brainstorm with ChatGPT

2. Research with Perplexity

3. Deep analysis with Claude

4. Visuals with Midjourney

5. Polish with Gemini

"Each tool adds unique value. Using just one is like cooking with only salt." - Kevin

Choosing the Right Tool
Decision Framework:

Need speed and creativity? → ChatGPT Need depth and accuracy? → Claude Need current information? → Gemini or Perplexity Need visual content? → Midjourney, DALL-E Need code help? → GitHub Copilot Need data analysis? → Julius

Real-World Tool Combinations

The Content Creator's Stack "I outline with ChatGPT, research with Perplexity, write with Claude, create images with Midjourney, and optimize for SEO with Gemini. What took a week now takes a day." - Content strategist

The Analyst's Toolkit "Julius for data exploration, ChatGPT for report writing, Claude for deep insights, Midjourney for visualization concepts. I'm not a team of one anymore – I'm a team of five." - Business analyst

The Educator's Suite "ChatGPT for lesson planning, Perplexity for fact-checking, Midjourney for visual aids, Synthesia for video content. My students get better materials than expensive textbooks provide." - High school teacher

The Cost-Benefit Analysis
Monthly Investment:

- ChatGPT Plus: $20
- Claude Pro: $20
- Perplexity Pro: $20
- Midjourney: $30
- Total: $90/month

Return: "That $90 saves me 20+ hours monthly. At my billing rate, that's $3,000+ value. The ROI is absurd." - Freelance consultant

Starting Your Multi-Tool Journey

Week 1: Master one tool deeply (start with ChatGPT)

Week 2: Add complementary tool (Claude for analysis or Midjourney for visuals)

Week 3: Integrate specialized tool for your biggest pain point

Week 4: Build workflows combining multiple tools

Month 2: Optimize tool combinations for your specific needs

Common Multi-Tool Mistakes

Tool Sprawl: Subscribing to everything, mastering nothing **Solution**: Add one tool at a time, cancel what you don't use

Redundancy: Using multiple tools for identical tasks **Solution**: Map each tool to specific use cases

Integration Failure: Tools in silos instead of workflows **Solution**: Design handoffs between tools

Cost Creep: Subscriptions without value tracking **Solution**: Monthly ROI review

Your AI Tool Audit

Answer honestly:

- ·What tasks take the most time?
- Where do you need higher quality?
- What capabilities do you lack?
- Which tools address these gaps?
- What's the learning curve?
- What's the potential ROI?

The Kevin Transformation

Six months later, Kevin leads his company's AI innovation team. "Starting with ChatGPT was perfect – it taught me what's possible. But discovering the ecosystem transformed how I work. I'm not just more productive; I'm capable of things I couldn't do before."

His advice: "Don't feel overwhelmed by options. Start with one, master it, then expand based on actual needs. The goal isn't to use all AI tools – it's to use the right ones for your challenges."

Your Tool Exploration Plan

This Week:

- List your three biggest work challenges
- Research which AI tools address each
- Try the free version of one new tool

This Month:

- Test one new tool thoroughly
- Compare it to your current solution
- Measure concrete improvements

This Quarter:

- Build integrated workflow
- Train team on successful tools
- Document ROI and benefits

The Future of Your AI Toolkit

"The best AI stack of 2025 might be completely different from 2024. The skill isn't picking the perfect tools – it's learning to evaluate and adopt new tools quickly." - Tech strategist

Key principles:

- Stay curious about new capabilities
- Test tools with real work challenges
- Measure results, not features
- Share discoveries with others
- Evolve continuously

Beyond Individual Tools

The future isn't about AI tools – it's about AI workflows. The most successful professionals don't just use multiple AIs; they create symphonies where each AI plays its perfect part.

Master this orchestration, and you don't just work faster – you work at a fundamentally higher level.

ChatGPT opened the door. Now it's time to explore the whole house. What room will you enter next?

Lesson 6.3: Where to Learn More – Continuing Your AI Journey

"I feel like I'm drinking from a fire hose," confessed Jennifer, a VP of Operations who'd been using ChatGPT for three months. "Every day there's something new. How do I keep up without it becoming a full-time job?"

I showed her my learning system – a curated approach that takes 30 minutes per week but keeps me at the cutting edge. Jennifer's eyes widened. "This is manageable. Actually, this is exciting."

The secret to continuous AI learning isn't consuming everything – it's consuming the right things in the right way. Let me show you how to build a sustainable learning system that evolves with you.

The Learning Hierarchy

Not all AI learning resources are created equal. Here's how to prioritize:

Level 1: Fundamentals (Master These First)

- Core concepts that don't change
- Ethical considerations
- Basic prompt engineering
- Understanding capabilities/limitations

Level 2: Practical Applications (Your Daily Focus)

- Use cases in your industry
- Workflow optimization
- Tool comparisons
- ROI strategies

Level 3: Emerging Trends (Weekly Check-in)

- New capabilities
- Industry disruptions
- Tool updates
- Future predictions

Level 4: Technical Deep Dives (As Needed)

- How AI actually works
- Advanced techniques
- Custom implementations
- Technical limitations

Your Curated Learning Sources
Daily Dose (5 minutes)
The Morning Scan:

1. AI News Aggregators:

- TheRundown.ai - Daily AI digest
- Ben's Bites - Conversational updates
- TLDR AI - Technical highlights

"I read one newsletter with coffee. Takes 5 minutes, keeps me current." - Marketing director

2. LinkedIn AI Voices: Follow 5-10 thought leaders in your industry using AI

- See real applications
- Learn from experiments
- Avoid echo chambers

Weekly Deep Dive (30 minutes)
The Friday Learning Session:
1. YouTube Channels:

- Matt Wolfe - Tool reviews and tutorials
- AI Explained - Deeper technical insights
- Fireship - Quick, dense updates

2. Podcast Rotation:

- Hard Fork - AI impact on society
- The AI Podcast - Industry interviews
- Practical AI - Implementation focus

"I listen during my commute. It's a learning time I already had." - Sales manager

Monthly Mastery (2 hours)

The Skill Building Session:

1. Online Courses:

- Coursera: "Prompt Engineering for ChatGPT"
- LinkedIn Learning: "AI for Business Leaders"
- DeepLearning.AI: "ChatGPT Prompt Engineering"

2. Hands-On Workshops:

- Local AI meetups
- Virtual workshops
- Industry-specific training

"One focused session monthly compounds into expertise." - HR director

Building Your Personal Learning Network

The Power of Community Learning

Online Communities:

1. Reddit:

- r/ChatGPT - Daily discoveries
- r/ArtificialIntelligence - Broader discussions
- r/LocalLLaMA - Technical advances

2. Discord Servers:

- OpenAI Discord - Official community

- Midjourney Discord - Visual AI
- Industry-specific servers

Real-World Impact: "I asked a question in Discord and got five solutions I'd never considered. Community learning is 10x faster than solo." - Developer

Professional Networks:

1. LinkedIn Groups:

- AI in Business
- ChatGPT for Professionals
- Industry + AI groups

2. Slack Communities:

- AI for Marketers
- GPT for Good
- Women in AI

The Experiment-Document-Share Cycle

Week 1: Experiment Try one new AI technique or tool

Week 2: Document Write up what worked, what didn't

Week 3: Share Post to LinkedIn or team Slack

Week 4: Iterate Improve based on feedback

"This cycle turned me from an AI consumer to an AI thought leader in my company." - Operations manager

Industry-Specific Learning Paths

For Business Leaders:

1. Start: "AI for Everyone" by Andrew Ng
2. Apply: HBR's AI articles for strategy
3. Connect: Executive AI roundtables
4. Lead: Create AI innovation committee

For Creatives:

1. Start: YouTube tutorials on AI tools
2. Apply: Daily creative experiments
3. Connect: Creative AI communities
4. Lead: Share AI-enhanced portfolio

For Technical Professionals:

1. Start: Fast.ai courses
2. Apply: Build something weekly
3. Connect: GitHub AI projects
4. Lead: Contribute to open source

For Educators:

1. Start: "Teaching with AI" resources
2. Apply: One AI lesson weekly
3. Connect: EdTech AI forums
4. Lead: Train other teachers

The Learning Tech Stack

Capture Tools:

- Notion/Obsidian for AI notes
- Loom for recording discoveries
- Screenshots for prompt successes

Organization System:

☒AI Learning/

├── Prompts That Work/

├── Tool Comparisons/

├── Industry Applications/

├── Future Trends/

└── Experiments Log/

☒Review Rhythm:

- Daily: Add one learning
- Weekly: Review and organize
- Monthly: Identify patterns
- Quarterly: Share major insights

Avoiding Learning Traps

Trap 1: The Shiny Object Syndrome Chasing every new tool without mastering any

Solution: One new tool per month maximum

Trap 2: The Theory Trap Learning about AI without using it

Solution: 80/20 rule - 80% doing, 20% learning

Trap 3: The Echo Chamber Only learning from similar perspectives

Solution: Deliberately seek contrarian views

Trap 4: The Overwhelm Spiral Trying to learn everything at once

Solution: Focus on immediate applicability

Your Personalized Curriculum

Month 1: Foundation

- Master ChatGPT deeply
- Read "Co-Intelligence" by Ethan Mollick
- Join one community
- Document 10 use cases

Month 2: Expansion

- Add one complementary tool
- Take one online course
- Attend virtual workshop
- Share learnings publicly

Month 3: Specialization

- Focus on industry applications
- Connect with practitioners
- Develop unique workflows
- Teach someone else

Ongoing: Evolution

- Weekly learning rhythm
- Monthly skill addition
- Quarterly strategy review
- Annual capability audit

Real-World Learning Success Stories

The Rapid Riser "I spent 30 minutes daily for 6 months learning AI. Got promoted to 'AI Innovation Lead' – a role that didn't exist before me." - Former analyst

The Career Pivoter "At 45, I thought I was too old for AI. YouTube University proved me wrong. Now I consult on AI implementation." - Former project manager

The Multiplier "I document everything I learn and share with my team. We're all growing together. Our department leads the company in AI adoption." - Team lead

The Meta-Learning Approach

Learn How to Learn AI:

1. Start with problems, not tools
2. Learn in public (share struggles too)
3. Teach to solidify understanding
4. Build learning into work, not on top
5. Focus on principles over features

Your Learning Action Plan

This Week:

1. Subscribe to one newsletter

2. Join one community

3. Try one new technique

4. Document the result

This Month:

1. Complete free online course

2. Attend one virtual event

3. Connect with 5 practitioners

4. Share 4 learnings

This Quarter:

1. Develop expertise in one area

2. Build network of 20+ AI users

3. Create learning resource for others

4. Establish learning rhythm

The Jennifer Transformation

Six months later, Jennifer leads her company's AI transformation. "The fire hose became a steady stream I could manage. More importantly, I learned to learn – that's the real superpower."

Her system:

- Morning: 5-minute newsletter

- Lunch: Browse AI community
- Friday: 30-minute deep dive
- Monthly: 2-hour skill session

"I spend less time learning than I used to spend worrying about falling behind."

The Only Learning Strategy That Matters

Here's the truth: The best learning resource is the one you'll actually use. The best learning schedule is the one you'll actually follow. The best learning method is the one that fits your life.

Build a system that works for you, not against you. Make learning inevitable, not optional.

The AI revolution isn't slowing down. But with the right learning approach, you don't need to keep up with everything – just everything that matters for your success.

Your future self will thank you for starting today. What will you learn first?

Lesson 6.4: Responsible Use Recap – Your Ethical AI Checklist

The email arrived at 3:47 PM on a Tuesday. "We need to talk about the AI incident," read the subject line from Legal.

Michael's stomach dropped. As team lead, he'd encouraged everyone to use ChatGPT for productivity. But someone had uploaded confidential client data, asked ChatGPT to analyze it, and now fragments were appearing in responses to other users. The breach wasn't malicious – just ignorant. But the consequences were the same: potential lawsuit, damaged trust, and careers at risk.

"If only we'd had clear guidelines," Michael told me later. "We were so focused on productivity gains, we forgot about responsibility. That mistake cost us a $400,000 settlement and our biggest client."

Michael's story isn't unique. But it is preventable. Let's build your ethical AI framework that enables innovation while protecting what matters.

The Ethical Foundation

Responsible AI use isn't about restriction – it's about sustainable innovation. Think of it like driving: traffic laws don't prevent you from reaching destinations, they ensure everyone arrives safely.

The Three Pillars of Ethical AI Use:

1. Transparency Be open about AI assistance

- Colleagues know when you're using AI
- Clients understand AI's role
- Work shows AI collaboration, not deception

2. Privacy Protect sensitive information

- Never upload confidential data
- Scrub identifying information
- Respect intellectual property

3. Accountability Own the outcomes

- Verify AI-generated content

- Take responsibility for errors
- Maintain human judgment

Your Ethical Decision Tree

When facing an AI use case, ask:

1. Is it legal?

- Does this comply with regulations?
- Am I respecting copyrights?
- Are there industry-specific rules?

If no → Stop If yes → Continue

2. Is it ethical?

- Would I be comfortable if this was public?
- Am I being transparent?
- Could this harm anyone?

If uncomfortable → Reconsider If comfortable → Continue

3. Is it professional?

- Does this maintain work quality?
- Am I adding human value?
- Will this build or break trust?

If breaks trust → Revise approach If builds trust → Proceed

4. Is it secure?

- Am I protecting sensitive data?

- Have I removed identifiers?

- Could this create vulnerabilities?

If risky → Add safeguards If secure → Implement

Real-World Ethical Scenarios

Scenario 1: The Resume Enhancement Question: "Can I use ChatGPT to improve my resume?"

Ethical approach: ✅ Use AI to improve formatting and clarity ✅ Let AI suggest stronger action verbs ✅ Have AI check for consistency ❌ Don't let AI invent experiences ❌ Don't claim AI's words as your writing style ❌ Don't fabricate skills or achievements

Scenario 2: The Client Proposal Question: "Can I use ChatGPT for client work?"

Ethical approach: ✅ Use AI for structure and ideas ✅ Disclose AI assistance if asked ✅ Add unique value and expertise ❌ Don't present AI work as fully human ❌ Don't use client data in prompts ❌ Don't rely on AI for critical decisions

Scenario 3: The Academic Assignment Question: "Can students use ChatGPT for homework?"

Ethical approach: ✅ Use for understanding concepts ✅ Let AI explain difficult topics ✅ Have AI suggest research directions ❌ Don't submit AI work as your own ❌ Don't bypass learning objectives ❌ Don't violate academic policies

The Data Protection Protocol

Before Using AI, Classify Your Data:

Public (Green Light)

- Published content
- General knowledge
- Public information Safe for AI use

Internal (Yellow Light)

- Company policies
- General procedures
- Non-sensitive data Remove identifiers first

Confidential (Red Light)

- Client information
- Financial data
- Personal details Never use with AI

Secret (Full Stop)

- Trade secrets
- Strategic plans
- Competitive intelligence Absolutely prohibited

Building Ethical Habits

Daily Practice:

- Pause before pasting into AI
- Ask: "Would I email this to a stranger?"

- Remove names, numbers, specifics
- Add value beyond AI output

Weekly Review:

- Audit your AI interactions
- Identify any close calls
- Update personal guidelines
- Share learnings with team

Monthly Check-in:

- Review industry AI policies
- Update security practices
- Assess ethical edge cases
- Refine decision framework

Common Ethical Pitfalls and Prevention

Pitfall 1: The Convenience Trap "It's faster to paste the whole document" Prevention: Create scrubbed templates

Pitfall 2: The Transparency Gap "They don't need to know I used AI" Prevention: Default to disclosure

Pitfall 3: The Ownership Confusion "AI wrote it, so it's mine" Prevention: Always add human value

Pitfall 4: The Verification Skip "AI is probably right" Prevention: Fact-check everything important

Your Personal Ethics Framework

Create Your AI Values Statement: "I use AI to _____ while always _____. I never _____ and I commit to _____."

Example: "I use AI to enhance my productivity while always maintaining transparency. I never compromise client confidentiality and I commit to verifying all important information."

Define Your Boundaries:

- What data will I never share?
- When will I always disclose?
- How will I verify outputs?
- Where will I add human value?

Build Your Safeguards:

- Technical: Password managers, encryption
- Procedural: Review before sending
- Cultural: Team agreements
- Personal: Ethical reminders

Industry-Specific Ethical Considerations

Healthcare: Patient privacy is paramount "We use AI for admin tasks, never patient data. Even anonymized data could be re-identified." - Hospital administrator

Legal: Privilege and confidentiality rule "AI helps with research and templates, never client specifics. We cite-check everything." - Law firm partner

Finance: Fiduciary duty guides decisions "AI analyzes public market data, never individual portfolios. Human judgment drives all recommendations." - Financial advisor

Education: Academic integrity matters "We teach students to use AI as a learning tool, not a homework completer. Citation of AI use is required." - Professor

The Reputation Equation

Your reputation = (Quality of work) × (Trust in process)

AI can multiply quality, but unethical use zeroes out trust. One breach can undo years of excellence.

Reputation Builders:

- Consistent transparency
- Higher quality through AI
- Teaching others responsible use
- Leading by example

Reputation Destroyers:

- Hidden AI use discovered
- Quality drops from over-reliance
- Data breach from carelessness
- Claiming AI work as sole creation

Your Implementation Roadmap

Week 1: Personal Foundation

- Write your AI values statement

- Audit current AI practices
- Identify risk areas
- Create data classification

Week 2: Practical Safeguards

- Build scrubbing templates
- Set up review processes
- Create disclosure language
- Test security measures

Week 3: Team Alignment

- Share ethical framework
- Discuss edge cases
- Agree on standards
- Document decisions

Week 4: Continuous Improvement

- Review near misses
- Update guidelines
- Share success stories
- Plan regular audits

The Michael Redemption

Two years after the incident, Michael's company has the industry's gold standard for ethical AI use. "That painful mistake became our greatest teacher. Now we innovate fearlessly within clear boundaries."

Their framework:

- Mandatory AI ethics training
- Clear data classification
- Regular security audits
- Transparent client communication
- Innovation within guidelines

"We're more productive than ever, and clients trust us more because we're open about our AI use. Ethics became our competitive advantage."

Your Ethical AI Pledge

Consider adopting this pledge:

"I commit to using AI as a tool for enhancement, not replacement of human judgment. I will:

- Protect confidential information
- Be transparent about AI assistance
- Verify important information
- Add unique human value
- Learn from mistakes
- Help others use AI ethically
- Evolve with technology responsibly"

The Bottom Line on Ethics

Ethical AI use isn't about being perfect – it's about being thoughtful. Every decision to use AI responsibly builds a foundation for sustainable innovation. Every shortcut risks everything you've built.

The companies and individuals thriving with AI aren't the ones pushing boundaries recklessly. They're the ones innovating responsibly within clear ethical frameworks.

Your ethical framework isn't a constraint on your AI use – it's what makes your AI use sustainable, scalable, and trustworthy.

Build it thoughtfully. Follow it consistently. Update it regularly. Your future success depends on the ethical foundation you lay today.

What ethical standard will you set?

www.ingramcontent.com/pod-product-compliance
Lightning Source LLC
LaVergne TN
LVHW051225080426
835513LV00016B/1423